MW00324787

For more information about the book,
plus coffee mugs and other merchandise,
see quietriverpress.com/friendly.html

THE FRIENDLY AUDIO GUIDE

by Mark Fleischmann

QUIET RIVER PRESS

NEW YORK

© Copyright 2018 Mark Fleischmann. All rights reserved.

No part of this book may be reproduced, stored in a retrieval system, or transmitted by any means, electronic, mechanical, photocopying, recording, or otherwise, without written permission from the author.

ISBN 13: 978-1-932732-20-7
ISBN 10: 1-932732-20-9

This book is printed on acid-free paper.

quietriverpress.com

For my mentors

Table of Contents

Introduction

Understanding audio technology shouldn't have to be hard. The purpose of this book is to make it easier. It is aimed at the beginner, the newbie, the person flustered by the jargon but determined to take advantage of the technology. Why bother? Here comes my mantra: To get *closer to music.*

Why do you need to understand the difference in bass response between a three-inch woofer and a five-inch woofer? To get closer to music. Why do you need to research amp specs? To get closer to music. Why do you need to delve into the arcane language of digitalia—the file formats, the bit depths, the sampling rates, all that messy talk? To get closer to music. Why upgrade your system in any way at all? To get closer to music, to improve your relationship with music.

The Friendly Audio Guide will differ in several ways from my previous technology book, *Practical Home Theater: A Guide to Video and Audio Systems.* This book is strictly for beginners, especially those putting together their first systems. It is deliberately short, just over 100 pages, and designed for people put off by having to wade through the copious detail of the other book. You should be able to get through most of it in one sitting.

Another difference is that this book is a pure audio guide. Unlike *Practical Home Theater*, it does not cover video and it places less emphasis on surround sound. Even so, buyers of the previous book should be warned that there is some overlap. My ways of explaining speaker and amplifier specs (among other things) haven't changed much, so some passages are repeated, most often in condensed form. I've tried to keep that to a minimum, but this book is intended for a different kind of reader.

In addition, this book will receive only occasional updates, if any; it will not be annually updated like the other book. If you've bought it once, you probably won't need to buy it again.

Finally, whereas the other book recommends only a smattering of products, this book recommends entire systems, mating speakers and amps at ascending levels of price and size. A webpage associated with the book will keep track of products going out of production and will recommend substitutes.

Why did I become an audio critic? Well, to get *closer to music* of course, but I got gentle pushes from friendships, magazine jobs, and serendipity. My introduction to real audio gear came from my Columbia classmate Henry Finke, who had a part-time job at a legendary store called Tech Hifi. The time we spent with his Infinity speakers and Crown amps showed me how much closer to music I could get. I wanted more; I was hooked.

My introduction to technology journalism came at *Video* magazine. I got that assistant editor's job not through the audiophile grapevine, but through a classified ad in *The New York Times*, like any other civilian. I seemed to have the balance of qualities the editors were looking for—interested in technology, but not consumed by it. I learned from the writers I edited.

After five years I knew enough to launch my writing career, forging relationships with enthusiast magazines such as *Audio Video Interiors*, *Home Theater*, and *Sound & Vision*. But my work has also appeared in general interest publications such as *Details*, *Rolling Stone*, and *The Village Voice*. I think of myself as a writer first and a tech expert second. As my editors at the music mag *Trouser Press* once put it: "If the records weren't free, we'd be rating cheeseburgers."

Though I started as an audio/video writer, I got a push toward the audio side from my first editor at *Rolling Stone*, Wook Kim, who offered me a chance to write about audio or video (but not both). I chose audio because my relationship with music is more intense than my relationship with movies. The magazine's name had magic properties in those days—it opened doors. I had only to ask for a review sample and it would appear on my doorstep.

To educate myself about high-end audio in particular, I found an early mentor in Steve Guttenberg, then a salesman at Sound by Singer, a much-respected audiophile retailer in New York. Many other folks in the audio industry have helped along the way, often quite selflessly. At press events and trade shows, I've found myself shaking hands and breaking bread with other people who wanted to get closer to music. It has not been a bad place to spend a writing career.

I write, you read. How should you cope with reading about audio? The audio press, once mostly limited to specialist magazines, has broadened (though not necessarily improved) to include web publications and forums. One unending debate that continues to play itself out is the objective approach vs. the subjective approach to audio equipment reviewing.

To oversimplify a little, objectivists say that the best way to evaluate audio gear is to measure it. The dean of the objectivist school was Julian Hirsch of *Stereo Review*, the pioneering audio critic who was among the first to measure audio gear with instruments. Objectivists are stout foes of hype and snake oil.

Subjectivists, on the other hand, insist that numbers don't tell the whole story, and that your own ears are the ultimate authority. Key figures in the subjectivist school have been J. Gordon Holt of *Stereophile* and Harry Pearson of *The Absolute Sound*. They redirected the path of audio criticism back to listening on a feeling level, exploring human sensations that escape the instruments and can't be quantified.

I write as a subjectivist but much of my work has been supplemented with measurements executed by others. As a practicing subjectivist, I depend on my ears, though as an informed reader, I have learned from both approaches.

The best way to evaluate audio gear is to listen. The best way to get closer to music is to listen. You get to do both kinds of listening at the same time. How could life possibly get better?

Putting Together a System

So you've decided to put together your first audio system (or your first serious one). Before you decide what to buy, think about what form your system should take. That in turn hinges on how and where you plan to use it. Are you a background listener or a foreground listener? The more casual background listener will have more leeway in system architecture and budget. Consider what would fit most easily into your life with minimal impact on your living space. The more active foreground listener will want the best system your budget will allow. The room will require components of a certain size, scale, and power. You will sit in the center sweet spot for maximum performance. If you are both a background listener and a foreground listener—depending on your mood or the time of day—then you'll probably want to put together a high-performing component system for serious listening in the main listening room and add simpler ancillary gear in other rooms for casual use.

Casual vs. component

The casual listener probably doesn't need the advice of a book. However, the audio industry sells a lot of casual-listening solutions, so let's run through them briefly. The simplest way to enjoy music is to plug *headphones* into your phone. Next in the pecking order is the *wireless powered* (or *self-powered* or *active*) *speaker*, generally Bluetooth and mono, though step-up models add other wireless technologies and stereo. But if you want stereo, the minimal spread of a one-piece speaker is too limiting, so step up to a *pair of powered speakers*, wireless or otherwise—then you can place them to achieve the best stereo spread and imaging of objects in the music. If your room demands more power, and your music library more connectivity, then you are ready for a component system.

In a *component system*, loudspeakers and amplifiers are separate. This moves the burden of equipment matching from the manufacturers to the buyer—so you'll need to study the chapters on speakers and amps to understand how their specs work. A component system should provide enough jacks to connect wired signal sources such as disc players, media players, and turntables. It may also need wireless protocols to connect your phone and tablet. Finally, a component system needs cables to lash it all together. You should budget enough for decent (though not necessarily expensive) speaker and interconnect cables.

The nerve center of your component system may be a *receiver, integrated amp*, or combination of *power amp* and *preamp* (together referred to as *separates*). These products come in stereo or surround versions. It is possible to buy an expensive receiver or an inexpensive power/preamp combo. But the most cost-effective systems are based on receivers and integrated amps, while the most powerful and costly systems are based on separates.

Getting and spending

How much should you spend on your system? Before you get down to numbers, first assess what you really need. You won't end up with the best system for your needs if you arbitrarily decide the total cost in advance. If you spend too little, you may end up with a system that won't make you happy—and that would just be a waste of money, putting you that many dollars further away from the system you really need. It is also possible to spend too much for a system you don't really understand, only to suffer buyer's remorse later. But here are a few suggestions:

If you're living in a studio apartment and your back is breaking under the weight of your college loan, buy the best pair of powered bookshelf speakers you can afford and run them with a wireless or wired connection from your phone or computer. With that strategy, two to five hundred bucks will go a long way. You can later add a disc player and/or turntable when you can afford to collect music in hard-copy formats. In the meantime, borrow CDs from the public library, rip them to a lossless format, and put a really good USB DAC (digital-to-analog converter) between laptop and system—you'll be amazed how much better your music sounds.

If you have real money to spend, and presumably a large listening room to fill with joy, you're in the market for high-end audio. Your speakers may be bigger, with more extended and powerful bass response, and your amp should support them with a wide dynamic range (the ratio between soft and loud). That in turn requires a good source of power, so look for beefy separates or an integrated amp with high-current power output. To make

The author in Shenzhen, China, where finding a store loaded with speakers is never a challenge.

those speakers and amps sound good, you will also need high-resolution digital or analog source components. In high-end audio, five thousand bucks buys a starter system and you can easily go into five or six figures.

If you fall somewhere in between, you are *the low end of the high end.* You want the best audio system your money will buy, and you're willing to spend a couple thousand bucks, but you want to stay out of nosebleed territory. You are going to have to work harder than Budget Gal or High-End Guy. If you live near an audio specialty retailer, or even a chain store with speaker displays, take advantage of the chance to audition a variety of gear. If you find what you need, buy it there so that the retailer who has served you will be around to serve you in the future. Otherwise the best way to forage for possibilities is to read, read, read. Since much of audio specialty retailing has moved out of local storefronts and onto the web, trying to get some sense out of Amazon user reviews is a necessary evil. But you may be surprised to find products in your price range carefully and thoughtfully reviewed in the high-end press.

Regardless of what you plan to spend, see the following chapters on each product category. And if you must buy on the web, get a money-back guarantee.

Loudspeakers

The amplifier may be the heart of your system but loudspeakers are the voice (and music is the soul). They turn audio signals delivered in the form of electric current into soundwaves. Some people refer to this process as "moving air" but it is more a matter of changing air-pressure patterns, like ripples in water. The bigger the speaker, the more air it moves. This section will talk a lot about stereo speakers and a little about surround speakers (which are covered in more depth in *Practical Home Theater*). It will start with the kind of speakers you connect to an amp, followed by powered speakers and wireless speakers, finishing up with a demystification of speaker specifications.

Stereo speakers

The earliest stereophonic technology dates back to the 19th century but stereo didn't become a mass-market product for the home until the late 1950s when the first stereo LPs and 45s became available. The advantage of stereo is that it creates a soundstage with voices and instruments *imaged* between the speakers. However, if the speakers are not of decent quality, fed by a good amp, and properly set up, imaging falters and there is a hole-in-the-middle effect. So you need to invest in good speakers.

The largest speakers are *floorstanding* or *tower* speakers. Their primary advantage is that their front baffles have enough room for larger drivers and/or more of them. That allows them to play deeper and more forceful bass without the assistance of a subwoofer. It also requires the receiver or amp to provide more juice. However, don't assume that all floorstanding speakers pro-duce deep bass—the frequency-response spec will provide a bet-ter indication of that. And don't assume that all floorstanders re-quire a lot of power from the amp. The sensitivity spec is the key to that. We'll get to those things later.

If you don't want big floorstanding speakers, consider what are variously known as *monitor, bookshelf,* or *stand-mount* speakers. They are still big enough to hold at least one large bass driver (5.25 inches and up) but small enough not to hog the room. In this size, you get adequate bass but not the deepest bass—some monitors need help from a subwoofer to reproduce the lowest note on a bass guitar and the full force of a kick drum. Note that so-called bookshelf speakers won't sound their best on a bookshelf—you'll get less congested bass and clearer midrange response if they sit on stands a foot or more from the wall. How-ever, this size may be the middle ground you're seeking.

The most compact speakers are *satellites*, which are almost invariably sold in *satellite/subwoofer* (or *sat/sub*) sets because their small drivers require the help of a subwoofer to produce even remotely adequate bass. However, they are helpful space savers, and that can make the difference between having a real component system or not having one. Satellites also make sur-round sound a more feasible option. They can be wall- or stand-mounted and the subwoofers that come with them are self-pow-ered (as subs generally are).

Surround speakers

The speaker categories above are sold for stereo systems but can also be used in surround systems. A surround system that includes towers will generally limit them to the front and right channels, though if you want an all-tower system, have at it. Monitors can be used for any channel. Unless you're running deep-voiced towers with one or more muscle amps, a surround

From left to right: A floorstanding or tower speaker, a monitor or bookshelf speaker, and a satellite speaker, shown in correct scale. The floorstander is the Paradigm Prestige 85F ($3998/pair), the monitor is the Paradigm Prestige 15B ($1598/pair), and the satellite is the Paradigm Cinema 100 CT ($999 for 5.1-channel system with subwoofer).

system will always include at least one subwoofer (or more than one for improved bass response).

Some speaker types are designed for specific uses in surround systems. The most prevalent one is the horizontal *center speaker*, which typically includes two large drivers flanking a small one. It's common—though not necessarily better—to use a horizontal center with towers and/or monitors in the four corners of the room. However, in a more ambitious system, it may be better to use perfectly matched speakers all around to preserve the integrity of the soundfield. If you must use a horizontal center, make sure it's *timbre-matched* with the other speakers. That means the speakers share the same tone, so that all of them reproduce objects in the same way, and none of them sticks out.

Other options in surround systems include dedicated *bipole/dipole speakers* in the rear of the room; they disperse sound more evenly and call attention to themselves less. *Height speakers* reproduce the height channels embedded in the new Dolby Atmos and DTS:X surround standards; they can be built into the room, or they can sit on your towers or monitors and bounce the height channels off the ceiling. The latter kind are called *Atmos-enabled speakers*.

A real surround system includes at least *5.1 channels*, meaning three speakers in the front, two in the side-surround positions, and the sub (the point-one). A *7.1-channel* system adds two back-surrounds. A *5.1.2-channel* system uses five floor speakers, one sub, and two height speakers. A *5.1.4-channel* system adds a second pair of height speakers. Variations abound.

Subwoofers

Subs produce nothing but low bass. They are used sometimes in stereo systems and almost always in surround systems. Nearly all subs are self-powered, even when combined with non-powered speakers. The smaller your speakers are—and the more limited their bass response, though that's not the same thing—the more they would benefit from the bass reinforcement of a sub. The sub's internal amp also frees the system from the power-sucking burden of reproducing deep bass.

A good sub can reproduce the deepest notes of musical instruments and the most aggressive movie effects. Adding a *second sub* provides more even bass coverage throughout the room; some surround receivers and preamp-processors make this easier with a second sub output.

Some of the best subs include *room correction* systems to improve irregular bass response. Even if you position your sub carefully and knowledgably, and even if you experiment with bass traps and other physical room treatments, many rooms have acoustic peaks or notches in the bass frequencies that only electronic room correction can fix. Some room correction systems, especially those built into surround receivers, correct all frequencies, in the speakers as well as in the sub—although those are more controversial because some of them introduce acoustic flaws of their own.

Soundbars and soundbases

The flatter TVs get, the worse their built-in speakers sound, so these self-contained audio systems are selling by the truckload. A *soundbar* goes in front of the TV or can be mounted on the wall below it. A *soundbase* actually supports the set itself.

A few distinctions to keep in mind: Bars and bases can be *active*, with built-in amps; or *passive*, requiring a receiver. They may have two, three, five, 5.1, or more channels—some even support height-enhanced Dolby Atmos. The sub may be built in or separate. If it's built in, the bass drivers will be small, making the name subwoofer more a marketing term than a real description.

Connectivity varies considerably. Most powered bars and bases connect through the TV—you connect all input sources to the

TV and an audio output cable from TV to bar/base. But some of the more ambitious ones have *HDMI* audio/video connectivity along with Dolby or DTS surround decoding, like a receiver. Non-powered bars and bases are fed by a receiver, acting as the three front channels in a surround system. Bluetooth is a plus.

Powered and wireless speakers

Most of the speaker categories above are *passive*, meaning powered by an amp. But if you don't want to base your system on amp, with all those cables, *powered speakers* (also called *self-powered speakers* or *active speakers*) are a simpler option.

Building the amp into the speaker allows the designer to tailor them perfectly (in theory). And you won't have to run cables from amp to speakers, though you'll still need cables to power the speakers' internal amps, to connect them to one another, and possibly to feed them with music. Powered speakers come in a wide range of sizes, from high-end powered towers to chunky monitors to compact one-piece wireless speakers.

Wireless speakers also come in a variety of sizes but most are small self-contained units. They can be either stereo or mono. Stereo models may have two channels, but when both channels are built into a single unit, the stereo spread is minimal—a pair of discrete wireless speakers has the sonic advantage.

The dominant wireless protocol is *Bluetooth*, a direct device-to-device connection that requires a simple pairing operation to connect to your phone or tablet. Bluetooth compresses the signal for distribution, on top of the compression also used in file formats such as MP3. Better units use higher-quality compression such as aptX or AAC to mitigate the damage, improving sound.

Some products use other wireless protocols that go through your home wi-fi network. These include Apple AirPlay, DLNA, DTS Play-Fi, Denon/Marantz HEOS, Sonos, Yamaha MusicCast, and WiSA. Some, like DLNA and Play-Fi, work with multiple brands; others, like HEOS and Sonos, are proprietary for those brands (though Sonos now licenses to Onkyo and others). These are not direct device-to-device connections like Bluetooth. All of them require a functioning wi-fi signal. But some of them do provide better sound and more versatility, such as the ability to chain multiple devices and to adjust settings with a phone or tablet app.

Interpreting speaker specs

A loudspeaker is a box, called an *enclosure* or *cabinet*, generally made of fiberboard, with dome- or cone-shaped *drivers* mounted on a front *baffle*. *Voice coils*, stimulated with current, move the drivers into action. A *crossover* routes the appropriate high or low frequencies to each driver. Those are the basics.

But how can you decode spec sheets to find what you need and get the best value for your money? A few absolutely indispensable specifications tell much of the story.

Frequency response reveals the range of frequencies (low, mid, and high sounds) and how evenly the speaker balances them. Manufacturers quote frequency response as a set of numbers that looks something like this: 40Hz-20kHz +/-3dB.

That may look like gibberish. But don't tune out. This is important stuff and not hard to understand. Let's break it down slowly, piece by piece.

Here's the executive summary: The first number is low-frequency or bass response. The second number is high-frequency or treble response. The third number is the loudness of each frequency.

And now here are more detailed explanations:

The first number is *low-frequency* or *bass response*, measured in Hz (Hertz). Speakers vary widely in bass response. While this roughly correlates with size—a satellite will produce far less bass than a floorstanding speaker—you can't assume that any two speakers of the same size have the same bass response. Specs matter here.

There is not much musically meaningful information below 20Hz, the bottom limit of most digital formats including the Compact Disc. Frequencies just above 20Hz capture most of the deepest impacts of a drum kit, especially the thud of the kick drum. A speaker with bass response of 40Hz and up will reproduce the bottom note of a bass guitar—that is the minimum you should expect in a speaker that claims to be musically self-sufficient, with no need for a subwoofer.

Bass frequencies down to 40Hz are musically meaningful. If they *roll off* (diminish) above that point, the lower notes in a bassline begin dropping out. Systems with monitor-size speakers and subwoofers usually assign frequencies of 80Hz and below to the sub. Systems with satellite-size speakers and subwoofers usually assign frequencies of 120Hz and below to the sub.

Let's move on to the second number. That is *high-frequency* or *treble response*, measured in kHz (kiloHertz). A speaker measuring up to 20kHz, the limit of most digital formats, reproduces sounds higher than you can hear (though your pets may hear them). Most speakers can manage that. The highest note of an 88-key piano is only 4.186 kHz. Most of what's above that is *air*, or ambience.

Finally we get to the *decibel (dB) range*. This is a measure of consistency in the balance of frequencies, which is especially important in assessing the musically crucial midrange. The lower the number, the better. The higher the dB number, the more the speaker is exaggerating or undercutting the music at certain frequencies, which can be especially noticeable with voices.

An overstated frequency, plus a certain number of dB over zero, is a *peak*; an understated frequency, minus a certain number of dB, is a *notch*. Within 3dB is reasonably consistent: the speaker has a few minor quirks, but they are not noticeable. Within 6dB is not as good: the speaker has some audible anomalies that become apparent over time. More than 6dB may be bad: falsifying texture, timbre, or bass weight. Within 1dB is heroic: the speaker has a consistently even balance of frequencies.

While you can understand frequency response with numbers, a chart makes it easier. A quick glance at the curve tells you much of what you need to know. Spec sheets usually don't provide charts, except for some high-end brands—but some audio mags so do in their reviews. Another advantage to independent measurements, as opposed to the official specs, is that they are less likely to include big fat barefaced lies.

Note that frequency response is usually measured *on-axis*, meaning in front of the speaker, dead center. But you won't necessarily be sitting directly in front of the speaker, so it can be helpful to see *off-axis* response as well, to understand how consistent the speaker sounds to those sitting off to the side, outside the sweet spot.

To sum up: Frequency response tells you how low the lows are, how high the highs are, how even the midrange is, and how all the frequencies are distributed and balanced (or unbalanced). Congratulations—you are now an expert in frequency response.

Another must-know spec is *sensitivity* or *room efficiency*, again specified in dB (decibels). This tells you how loud the speaker will play overall. It is usually measured with a one-watt test tone one meter from the speaker. Sensitivity of about 88dB is average. Lower than that will require a more powerful amp. Higher than that will run well with a more modest amp.

Sensitivity is measured in an *anechoic* (non-echoing) test chamber. When measured in a room, it is called room efficiency, and you can assume rating inflation of a couple of dB. Never forget that every 3dB cut in sensitivity requires a doubling of amp power to achieve the same volume. You may see a *power handling* spec saying how much power the speaker needs from the amp—but sensitivity and room efficiency are more meaningful.

Impedance is another spec that relates to the amp. It is the amount of resistance a speaker puts up to an audio signal entering through its terminals. Impedance is like a faucet. When impedance drops, the faucet is opened and more current runs through, which means more work for the amp. An average speaker impedance is 8 Ohms, which is easy for any amp to drive. A 6 Ohm speaker is a little harder to run, 4 Ohms harder still, and 2 Ohms demands a very beefy amp. Some amps are switchable for different impedances.

If you're buying speakers to use with a low- to midpriced receiver or a sweet little tube amp, 8 Ohms is the magic number. Speakers actually operate at a range of impedances, but are usually specified at a single *nominal impedance*. Impedance is also a part of amplifier specs, and is helpful in achieving a good mating of speakers and amps.

Then there are the *drivers*. A *two-way speaker* combines a *tweeter*, which reproduces high and mid frequencies, with a *woofer*, which reproduces mid and low frequencies. A *three-way speaker* adds a *midrange* driver to reproduce midrange frequencies, including much of the human voice.

Driver sizes and materials can have a strong impact on how a speaker sounds. Tweeters are generally inch-wide domes. The larger the woofer cone, the more bass the speaker can reproduce. A speaker used without a subwoofer generally requires a woofer size of 5.25 inches and up—although that hinges in part on how the designer decides to build it. A woofer attached to the baffle with a more flexible high-excursion *surround* can move farther and produce more bass. Textile tweeters often have a softer sound than aluminum tweeters, though that may be a question of voicing—in other words, how the designer uses it to produce a desired sound.

The *speaker enclosure*, the box itself, can also affect sound. Most speaker cabinets are made of medium-density fiberboard though cheaper ones may be plastic. The thicker, the better. A thicker enclosure is less prone to coloration and sound-polluting resonance. A good enclosure may also include internal bracing or multiple layers to further tame vibrations in cabinet walls.

The enclosure will usually be called *ported, vented,* or *bass reflex*—in other words, it has a big hole in back or front to let more bass into the room. This can also result in a bass peak and muddier bass, depending on the designer's skill and your choice of speaker placement. A *sealed* or *acoustic suspension* enclosure dodges that problem by eliminating the port. That is slightly less efficient but does give you more placement options, since you won't have to worry about the audible impact of the port.

Speaker terminals come in two basic types. The better one is the *binding post*, which accepts banana plugs, spade lugs, pin connectors, or bare cable tips. Screw-down binding posts are more versatile than the spring-loaded binding posts found on some compact speakers. The alternative is the spring-loaded *wire clip*, which has a little pull-down tab and works best with bare cable. Better speakers, or any speakers that aspire to be better, have binding posts. Cheap ones may have wire clips. Avoid buying those.

Amplifiers

The heart of your system is the amplifier. In a two-channel system it may take the form of a stereo receiver, stereo integrated amp, or combination of stereo preamp and power amp. In a multi-channel system it may take the form of a surround receiver or a combination of surround preamp-processor and multi-channel power amp.

A *stereo receiver* includes an amplifier section to power your speakers, a preamp section to handle source switching and volume control, and an AM/FM tuner. A *stereo integrated amp* is similar to a stereo receiver but eliminates the tuner to avoid the noise associated with it; it is also marketed with a different attitude, for the more discriminating listener. In the final step up, the *stereo preamp* and *power amp* are separated—hence the term *separates*—to protect the preamp from being polluted by noise from the high-current power amp. A *stereo-separates* system may pair the preamp with either a single stereo power amp or two mono power amps for even more power and control.

A *surround receiver* (or *A/V receiver*) adds many features to the basic configuration of the stereo receiver. They always include five or more amp channels, surround processing, HDMI audio/video switching, HD or Ultra HD video processing, and remote control; sometimes they also support room-acoustic correction, wireless audio, network audio, multi-room audio, and app or voice control. Surround preamps are usually called *preamp-processors* (or *pre-pros*) because they include the surround processing necessary for surround sound; the power amps in a surround-separates system may range from a single multi-channel unit to a rack full of multi-channel, stereo, or mono amps. Separates are generally more powerful and more expensive than receivers or integrated amps, with certain exceptions. Pardon the incessant plugs, but surround electronics are more thoroughly covered in my other book *Practical Home Theater*.

Note that you can buy a surround receiver for the extra features—wireless connectivity, room correction, and app control, for instance—while connecting just a single pair of speakers. The receiver's setup menu allows you to specify two-channel operation—or 2.1-channel, if you want to add a subwoofer to beef up the bass. There is no law saying you have to buy five or more speakers and a subwoofer to go with a surround receiver. On the other hand, certain stereo-oriented audiophiles would like to remind you that surround receivers are loaded with features you may never use, and each of those features is a line item in the design budget that diverts resources from things that would enable better sound quality. Whatever choice you make, the audio police will not break into your home and bust you for crimes against the state.

Tube, solid-state, and vintage amps

Audio started with *tube amps*, which use vacuum tubes to amplify the signal, and progressed to *solid-state amps*, after the invention of the transistor. Most amps are solid-state amps. Some audiophiles (and other hobbyists) seek out tube amps for their warmer, if not necessarily more accurate, sound.

There are reasons why tube gave way to solid state. Tube amps are more subject to distortion—though it's a more benign kind of distortion that affects a different set of harmonics, and some feel it adds to the sound. Tubes also morph throughout their lifespans, changing the sound, and eventually need replacement; solid-state amps also age but more slowly and gradually. Some audiophiles love swapping out tubes to change the sound of their amps, a practice called *tube rolling*.

Tube amps are more suitable for stereo systems than for surround systems with their extra channels. Though some tube-based systems can be quite powerful, lower-powered tube amps can produce a lovely sound when mated to suitably efficient speakers. Tube gear can be very cheap, for a no-name product sold on Amazon, or very costly, for the best separates from high-end tube specialists like Audio Research and Conrad Johnson.

You may be tempted to buy a used tube or solid-state amp to get a vintage golden sound. That can be a valid option if you know what you're doing. However, you must be willing to accept the risk, including the lack of warranty repair. Obviously, a vintage tube amp may need new tubes, and they may cost a bundle. Less obviously, a solid-state amp may need to be *recapped*—replacing the power capacitors, whose electrolyte fluids dry out over long periods of time. Any kind of vintage audio gear would benefit from a good internal cleaning, including the de-oxidizing of controls, so factor the cost of service into your budget if the seller has not already done the renovations.

Interpreting amp specs

How can you read amplifier or receiver spec sheets to find the product best for you? *Power ratings* are crucial. Seek out manufacturer claims in spec sheets and independent confirmation in reviews.

A typical power rating may read: "100 watts per channel, RMS, 20Hz-20kHz, into 8 Ohms, at 0.5% THD." Based on specifications defined by the Federal Trade Commission, that means the amp delivers 100 watts to each speaker, with continuous power, at the full range of frequencies, and that it mates well with speakers having a *nominal impedance* of 8 Ohms (see the discussion of speaker impedance specs in the previous chapter). *THD*, or *total harmonic distortion*, measures the amount of impurity introduced by the amp.

The greater its workload—depending on content, speakers, and volume setting—the more an amp distorts. Accelerated, audible distortion is called *clipping*. A powerful amp can put out its full specified power before distortion rises to an objectionable level. A less powerful one distorts sooner and sometimes more harshly. Note that distortion is a question of quality as well as quantity. Vacuum-tube amps may have lower rated power and higher rated distortion. However, they often distort in a way that is easier to listen to, and even pleasant. Nasty distortion affects higher-order *harmonics* (the overtones that accompany a basic pitch). A well-designed tube amp distorts in a more listenable manner, affecting lower-order harmonics. If you're buying a solid-state amp, pay attention to those THD numbers, though small differences may not be meaningful.

Power ratings may take different forms from the FTC method described above. Some are measured with a *1kHz test tone* as opposed to a full range of frequencies. This provides a stable frame of reference but one that has little to do with music. Also, some power ratings may specify *peak power*, measuring the amp's performance at a loud, transient high point—as opposed to the more demanding *continuous power*. That often happens with subwoofer amp specs, which are exempt from FTC rules. Amp specs are most useful when they require all channels to be driven at the full range of frequencies at the rated wattage.

There are lies, damned lies, and amplifier specs. How do manufacturers cheat? They might narrow the bass response to eliminate the power demands of low-bass reproduction. Or they might run just one channel at a time for stereo amps or (very commonly) just two channels at a time for surround amps. Specifying an impedance lower than 8 Ohms also drives up power ratings (because a lower-impedance speaker accepts more current). However, a manufacturer that specifies power ratings into multiple impedances is providing useful information. Two different models may both be rated at 100 watts into 8 Ohms, but if one delivers 120 watts into 6 Ohms while the other delivers 150 watts into 6 Ohms, the latter delivers more current.

All other things being equal, more current is better. It makes your system play louder, sound cleaner, and less likely to fry the speakers with distortion. An old audiophile adage says that you can never have too much power while you can certainly have too little power. Note that identical power specs don't necessarily mean identical sound. A good amp is more likely to have a heavier power supply and high current capability, so check those weight specs.

Amplifier topology takes a few different forms, each of which has a different impact on sound. *Class A*, the traditional though arguably obsolete high-end choice, keeps both of its power stages full of current, which sounds great but does dissipate a lot of power in the form of heat. Rarely used *Class B* pushes and pulls current between the two power stages, so that one is always on and the other off. Most analog amps are a *Class AB* hybrid which keeps current in both stages part of the time but alternates between them, providing the ideal compromise of sound quality, efficiency, and cost. *Class D* amps, sometimes incorrectly referred to as digital amps, convert the signal to a train of pulses that is always on or off. This is more energy efficient and allows amps to be built smaller and lighter. Any of these can sound great or terrible, depending on design and build.

Wireless features

As explained in the last chapter's discussion of wireless speakers, there are two kinds of wireless connectivity you might find in receivers and amps. The dominant one is Bluetooth, a device-to-device connection that can benefit from higher-quality compression such as aptX or AAC. The alternatives go through your home wi-fi network. Being app-driven makes them more versatile and adjustable.

Some wireless schemes are built into certain brands of product—for example, HEOS in Denon and Marantz receivers, or MusicCast in Yamaha receivers. Step-up surround receivers offer built-in support for music streaming services such as Spotify, Tidal, and internet radio via TuneIn.

However, if you want to prioritize your amp's sound quality over its wireless features, you might buy the best-sounding amp you can afford and provide for wireless through inexpensive add-ons. Some receiver makers that don't include Bluetooth as a feature offer an optional dongle. To support AirPlay, Apple offers the AirPort Express. Google's ChromeCast supports its own music services, Google Play and YouTube, as well as the other popular streamers mentioned above. Roku also supports numerous music channels. Note that Chromecast and Roku are limited to the HDMI interface found in audiovisual gear.

You may also get network-delivered audio into your system through various kinds of media player, in addition to the dongles mentioned above. The chapter on digital sources will go into more detail on that kind of component.

Control and ease of use

One advantage of buying a receiver is that you're almost guaranteed to get a remote control. Don't assume that anything you buy will have a remote. Stereo and surround receivers have them, as do surround pre-pros, but some stereo integrated amps and preamps lack them. Remotes that come with surround receivers may also be programmed to operate other components such as TVs and disc players. Products that use the HDMI interface—such as surround receivers, Blu-ray players, and TVs—can also coordinate power-on via *CEC*, or *consumer electronics control*. Go into the TV's menu to "discover" other CEC components.

In surround receivers and pre-pros, the remote control meshes with an onscreen *user interface* to provide access to the product's many features and functions. However, that is not the only (or necessarily the best) way to control the beast. Most major surround receiver makers offer a *phone or tablet control app* that liberates you from the TV-screen interface.

Some audio products (such as soundbars) are just starting to offer *voice control* via Amazon's Alexa or the Google Assistant, among others. This technology is in its infancy and does not work perfectly in my experience. If I instruct Alexa to change inputs on my TV, she is fussy about the terminology, and even if she agrees to do it, she does not always succeed. But voice control is unstoppable and likely to improve.

Ease of use is an issue that often emerges after you've invested in an audio product—so try to anticipate it before you whip out the plastic. Are the front panel controls easy to find and legibly labeled? Are the remote-control buttons well differentiated by size, shape, color, and layout? Is the app well-designed? In a surround receiver, is the onscreen user interface legible and easy to navigate (inasmuch as these things are ever easy)?

Connectivity

The back panel of a receiver or other audio product may not be the most appealing thing to contemplate. But it has a major impact on what you can connect to the product and how you will use it.

There are *high-level signals* and *low-level signals*. *Speaker cables* carry high-level (or *speaker-level*) signals. *Interconnect cables*, analog or digital, carry low-level (or *line-level*) signals. Your system needs both to function.

Any receiver, integrated amp, or power amp will have *speaker terminals*. These usually take the form of *binding posts*. The best binding posts are the *five-way* kind, which can accept *banana plugs, double banana plugs, spade lugs, pin connectors,* or *bare wire*. Surround receivers usually have *UL-approved*

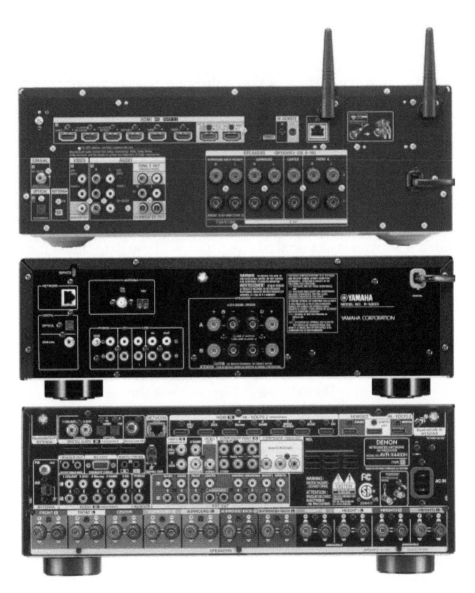

How much connectivity does your system need? The Yamaha stereo receiver shown at center accepts three analog sources plus a turntable. The Sony budget surround receiver at top adds several HDMI jacks for A/V sources. The Denon surround receiver on the bottom adds more of everything, though the main reason to buy it might be more power and/or features.

collared binding posts which prevent the use of spade lugs. That is not a problem for reviewers like me who prefer the convenience of banana plugs. I've observed banana plugs in use in facilities where some of the world's leading loudspeaker designers do their research and development. Beware of cheap products that offer spring-loaded *wire clips*, which are fragile, less versatile, and do not provide a secure connection.

Interconnect cables tether turntables, disc players, and other source components to the receiver, integrated amp, or preamp. *Analog interconnects* usually use two of the familiar RCA-type plugs, color-coded red and white, one for each channel. A turntable will require a special *phono input*, and the latter must mate with a specific kind of phono cartridge, either moving-magnet or moving-coil. *Digital interconnects* may be either *coaxial* (using a single RCA plug) or *optical* (using a fiber-optic cable and plug). Most fiber-optic cables are plastic fiber (*Toslink*) though some are glass fiber. Many audiophiles prefer coaxial to optical, citing bandwidth and durability. Optical cables are less prone to hum, though they are also extremely fragile. Kink the cable, or bend it too sharply, and it stops working.

Stereo and surround separates require analog interconnect cables between the preamp (or pre-pro) and power amp(s). The three-pin *balanced XLR* connection is best, especially when used over long distances—for instance, to link a preamp on a rack to heavy amps on the floor. The two-pin *unbalanced* or *single-ended RCA* connection is next best. It works acceptably well over short distances.

Surround receivers are heavily dependent on the *HDMI* interface, which carries both video and various forms of surround sound. Don't buy an old pre-HDMI receiver—it may not be able to feed a modern flat-panel video display with HDMI inputs. Also be aware that older versions of HDMI carry less advanced forms of surround sound. To make a long story short, when buying surround products, buy recent ones.

What kind of cable is best for your system? Are premium cables worth the extra cost? The chapter on cables will discuss them in more depth.

Digital Sources

It can be tempting to think of your system as a mating of speakers and amps and deprioritize the rest. But source components can have a noticeable impact on sound. It doesn't make any sense to buy good speakers and amps and feed them garbage. If you can't afford good source components right away, at least keep them mind as system upgrades. The good news is that there are inexpensive ways to get good digital sources into your system. Let's go over the options for disc players, digital-to-analog converters, streaming, downloading, ripping, and media players.

Disc players

Why bother having a disc player in your system? If you don't plan to go beyond streaming and downloading, skip to the next section. But if your local ISP and network connection are unreliable, a hard-copy disc library is a must for those times when you need music and don't want it hear it glitching. A disc library might be something you inherit or something you build (see chapter on collecting a music library).

The *Compact Disc format* is showing its age but there are still valid reasons to collect CDs. You might want to hear music unavailable from the streaming and download services. You might discover that the MP3 album you found on Amazon is actually more expensive than the CD. Your well-meaning aunt may give a CD as a birthday gift.

CD-only players are still sold but they have gone from mass-market to specialty items, sometimes with bleeding-edge prices. It is more pragmatic to buy a *Blu-ray player*—not only to play

Blu-ray discs and DVDs, but to play CDs. A Blu-ray player is designed to work primarily with the HDMI interface in an audio/video system. To access its menus, you would need to connect a TV. But if you don't want a TV in your system, the player and its remote will still have enough basic transport controls to spin CDs without access to the menus.

Another option is the *universal disc player*, which plays Blu-ray, DVD, CD, and adds the dueling high-resolution audio formats *DVD-Audio* and *Super Audio CD*. DVD-A and SACD are music formats that support surround sound and high-resolution (better than CD) audio. Some call them "dead" but if you're into classical music or progressive rock, there is still a steady trickle of releases. However, the Blu-ray format is also becoming a "pure audio" format thanks to its support for surround and high-res. That probably means a diminishing future for DVD-A and SACD. Note that DVD-Audio discs come with both a high-res MLP soundtrack and a more compatible Dolby Digital soundtrack which can be decoded by any surround receiver. SACDs usually (though not invariably) come in the form of hybrid discs with a high-res and surround-capable SACD layer and a lower-res CD layer, which allows them to be played on any kind of disc player and to be ripped for use on your computer, tablet, or phone.

DAC magic

The best way to improve the quality of digital audio in any system is to use a high-quality *DAC*, or *digital-to-analog converter*. If you are assuming any digital source component will be good at this crucial act of conversion, you are mistaken. Having a DAC isn't the same as having a good DAC.

Whether you should worry about the DAC depends on the source component and the architecture of your system. If your digital sources are feeding a stereo or surround system via digital interfaces (HDMI, coaxial, optical) then the conversion is being handled by your receiver, integrated amp, or preamp. If you are using a high-end disc player, it is probably using a good DAC to serve its analog output (but you are probably not using a high-end disc player). When you really have to be pro-active about getting a decent DAC into your system is when you are using a computer, tablet, or phone to play or stream music via the USB interface.

A computer is a chamber of horrors for an audio signal. It is a big box of interference that muddies the signal. And its constant multi-tasking causes audible degradation called *jitter*, bringing chaos to the delicate timing of digital audio's stream of zeroes and ones. Your computer has an analog audio output backed by a cheap DAC in its soundcard but don't use that piece of junk. Instead, insert a USB DAC between your computer and one of your audio system's analog inputs. It will avoid much of the interference and bypass the computer's clock, bringing order to the timing of the zeroes and ones. This will result in cleaner, finer-grained, more listenable sound.

An adequate DAC can cost as little as $100 and may be no larger than a USB stick drive. Pricier DACs—ranging in size from small boxes to full rack-size components—can provide even better sound and more connectivity options including digital outputs. DACs can also double as *headphone amps*, not only cleaning up the signal, but amplifying it and adding volume control. Most headphones sold today are efficient enough to run off tablets or phones, but more demanding ones benefit from the extra juice provided by a DAC-amp. To adapt a USB DAC to a tablet or phone, use an *OTG adapter* for Android gear or an *Apple camera adapter* for iOS gear.

Streaming

Digital culture has a way of undermining tyranny. Concentration of ownership and rigid program directors may have destroyed commercial radio—but streaming has become the new radio. It has also overtaken both discs and downloads in sales revenue.

Music streaming is most commonly done via computer, tablet, or phone. But support for music streaming is also built into audio/video dongles like Google Chromecast, boxes like Roku or Apple TV, surround receivers, and TVs.

Depending on the service, you may be able to choose from free, premium, family, or high-resolution tiers. Some services also offer discounted student and military tiers. Free service is usually supported by advertising (like commercial radio). You might get a half-dozen songs before the ads kick in. To eliminate the ads, support multiple accounts, or access high-resolution audio, step up to higher tiers.

For the largest library, try Spotify. To set up personalized "radio stations" that follow a musical train of thought, try Pandora. Tidal has become an audiophile favorite—not only because it offers a CD-quality lossless "Hi-Fi" tier, but also for its MQA-compatible "Master" tier, which supports a novel new technology that fits high-res quality into a CD-quality-size stream. Apple and Amazon offer services designed for their ecosystems—tyranny will always find a way to make a comeback.

Downloads and ripping

Streaming has now overtaken downloading in sales revenue. But the advantage of downloading is that you own the audio files forever, or for as long as you manage to keep them stored. And you're not at the mercy of your internet connection. The music is there when you want it.

Downloading also enables you to choose the audio file format in a more specific way than streaming. That is significant if sound quality matters to you. The same content sold in an inferior lossy file format on Amazon or iTunes is also sold in a superior lossless or uncompressed file format on HDtracks or Acoustic Sounds.

Downloading does have downsides. Your library remains accessible only for as long as you successfully keep the file stored, so be careful to back it up onto multiple hard drives, other devices, or cloud storage to avoid a catastrophic crash. Downloads usually don't come with cover art, except for a stingy little thumbnail that may or may not show up on your playback device. And they don't come with liner notes—LPs and CDs have the (potential) advantage there. The biggest downside is that your download library, like your disc library, is only as big as you can afford to make it (unless you steal). You won't get the incredible selection you'd get from the likes of Spotify.

However, downloads can be shared among friends (if not hampered by anti-copy restrictions). And downloading is not the only way to acquire audio files. You might also rip your CD collection or borrow CDs to rip. If you do rip CDs, be sure to rip them in a lossless format such as FLAC or ALAC to avoid reducing sound quality.

Audio file formats

That brings us to the subject of audio file formats. There are three basic kinds: lossy, lossless, and uncompressed.

Lossy formats are the most efficient kind but that efficiency comes at a price. They use *perceptual coding* and *data reduction* to prioritize some kinds of data over others, eliminating about 80 percent of the original data. The lower the *bitrate*, the smaller the file, but the more obvious the degradation. A file ripped from CD at 320 kilobits per second will sound closer to the original disc than one ripped at 128 kbps—the difference is very audible to most listeners. So if you are buying or ripping in a lossy format, go for the highest possible bitrate. Examples of lossy formats include Amazon-approved but generic MP3, iTunes-approved AAC, Microsoft-approved WMA, and open-source Ogg Vorbis.

Lossless formats also use perceptual coding and data reduction but eliminate less data, about 50 percent, packing and unpacking the data in a clever way that allows the original bitstream to be reconstructed bit for bit. This is the best compromise between efficiency and sound quality. Lossless formats include open-source FLAC, Apple-approved ALAC, Microsoft-approved WMA Lossless, and the MLP format used on DVD-Audio discs.

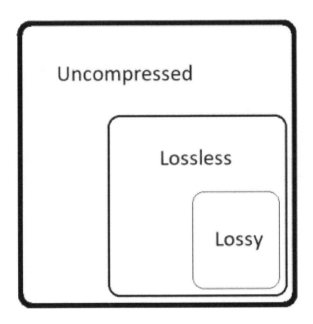

Uncompressed file formats include all the bits in the original signal, omitting nothing. Lossless file formats omit some data but still permit a bit-for-bit reconstruction of the original signal on playback. Lossy file formats are the lowest-quality kind. They omit a large percentage of the original data and do not permit all of it to be recovered on playback.

Uncompressed formats, unlike lossy or lossless, do not eliminate any data. They are used by the recording industry though you can also download uncompressed audio if you know what you're looking for or rip it from CD. Examples include Microsoft's WAV format, Apple's AIFF format, and audiophile-approved DSD.

The generic name *PCM* (*pulse code modulation*) is given to CD audio as well any other kind with a bit depth and a sampling rate. What are bit depth and sampling rate? They bring us to the next discussion. Be forewarned that the level of difficulty is about to rise. Proceed at your own risk.

Digital audio basics

If hearing about the guts of digital audio would make you flee the room screaming, skip this section. But to make more informed downloading and ripping decisions, it helps to know a little about digital audio basics.

Digital audio is basically a *string* of zeroes and ones. They are called *bits*. *Bit depth* refers to the number of zeroes or ones in a digital string. *Sampling rate* (or *bitrate* or *data rate*) is the number of strings transmitted per second.

For example, CD audio is a 16-bit, 44.1 kHz (or 16/44.1) format. That means it transmits strings of 16 zeroes and ones 44,100 times per second. To provide a contrasting example, one form of high-resolution audio is 24-bit and 96 kHz (or 24/96). It transmits strings of 24 zeroes and ones 96,000 times per second.

The higher the bit depth, and the higher the sampling rate, the better the sound. The 24/96 high-res file in our example has the potential to sound better than the 16/44.1 CD-quality file—if both are based on the same master, and if the underlying recording is a good one. And the 16/44.1 file will sound better in its uncompressed or lossless form than in its degraded lossy form, with most of the data thrown away.

Here is a rough analogy: Bit depth and sampling rate are like a wine bottle. Audio file formats such as MP3 or FLAC are like the amount of wine in the bottle. And the original source recording is akin to the quality of the wine in the bottle. You want the biggest bottle with the greatest amount of the best stuff.

You might also think of this as a pecking order. A high-quality analog or digital recording, delivered at 24/96, in lossless FLAC, has the potential to sound great. The same recording, delivered in CD-quality 16/44.1, has the potential to sound somewhere between great and very good. The same CD, ripped to MP3 at 320 kilobits per second, is not quite as good. The same CD, ripped to MP3 at 128 kbps, will suffer degradation most people can hear.

The best-case scenario is a high-quality analog or digital recording, encoded at the highest possible bit depth and sampling rate, stored or transmitted in an uncompressed or lossless file format. The worst-case scenario is a low-quality recording, encoded at lower bit depth and sampling rate, transmitted in a lossy file format. There are many variations in between.

Caveat emptor: Let the buyer beware.

MQA: new high-end contender

MQA, or *Master Quality Authenticated,* is a new form of audio encoding that breaks many of the rules mentioned above. It uses a complex process (just in case digital audio wasn't thorny enough) nicknamed *audio origami* to fold down bulky high-resolution audio into smaller streams or file sizes. It may well improve the sound quality of downloads, but its greatest potential is in streaming, because it enables high-resolution audio to travel in smaller CD-quality streams. Tidal is the first streaming service to offer it in meaningful amounts.

Critics of high-res audio in general, and MQA in particular, say human ears don't need ultrasonic frequencies and extreme dynamic range for music to sound good. But the developers of MQA are more concerned with tiny differences in signal timing that they contend are embedded into our deepest survival responses, according to the latest psychoacoustic research. Another advantage of MQA is the potential to correct various distortions introduced by all the stages of gear between the recording studio and your system.

You can stream or download MQA-encoded material and decode it with DACs (digital-to-analog converters) that cost anywhere from a hundred bucks to the stratosphere.

Component media players

Consumer-level digital audio started with the venerable CD player and has evolved to embrace mobile devices and the numerous formats and technologies described above. The component *media player* (or *network audio player*) brings it back to where it started, with a rack-size product that makes it easy to enjoy the magic of those zeroes and ones.

Why use a hardware-based media player in lieu a software-based media player running on your computer or mobile device? Maybe you don't want your system to be tethered to a desk, especially if you've been sitting at a desk all day. Maybe computer audio is annoyingly complicated (die, iTunes, die). Maybe you'd like something that fits into a rack better than a laptop or tablet. Maybe you're suffering from smartphone fatigue. Maybe you'd like to keep the smartphone free for other uses. Or maybe you'd prefer to use a traditional remote control—though media players, like receivers, can run off tablet or phone apps.

Media players can serve up streamed content, stored content, or possibly both. If you are addicted to a particular streaming service, make sure it is supported unless you want to deal with a box or dongle. A media player with a *hard drive*—making it more of a *media server*—can store and access your library. The bigger the hard drive, the more it can store. You might prefer a media player that pulls media off an *external hard drive*, which can be upgraded at will. One with a *CD drive* can help with the ripping.

A really good media player will have an *interface* that makes it easy to navigate your library and find what you want to hear. And it will have a high-quality *DAC*, because it would be no fun to hear your music botched by low-quality digital-to-analog conversion. Sound is inherently an analog creature. A media player, like any other audio component, has to get those analog soundwaves to your speakers and your ears.

Turntables

The renaissance of the LP is just the latest chapter in the annals of analog playback, which goes all the way back the dawn of audio history with cylinder recordings and 78s. The 12-inch, 33-rpm LP format as we know it was introduced in mono by Columbia Records in 1948 and went stereo in 1958. From then to the mid-1980s, the LP—and its little sister, the 7-inch, 45-rpm single, introduced by RCA in 1949—were the dominant music formats. It's generally forgotten that the analog audiocassette was first to briefly replace the LP as the biggest-selling music format.

But was the introduction, by Sony and Philips, of the first mass-market Compact Disc players in 1983 that signaled the LP's death knell—or so it seemed. The digital CD's "perfect sound forever" made the analog LP obsolete, consumers were told by reputable critics and publications. Almost overnight LPs disappeared from record stores, to be replaced by shiny five-inch discs in awkward longboxes at twice the price. (At the moment of the format's supposed demise, a new LP cost $7.) Most music lovers dumped LP collections to repurchase their favorites on CD and the music industry made a ton of money—but it was also a great time to build a used-LP library for next to nothing.

What followed was the greatest comeback story in the history of consumer electronics. A cadre of audiophiles and other analog loyalists held on, insisting that the CD's virtues were oversold and unconscionably hyped. A fierce (and ongoing) digital-vs.-analog discussion ensued. Even as LP sales hit bottom, the LP started to acquire new credibility as an underground niche format. Some manufacturers continued producing turntables. A few of those manufacturers were born at virtually the same moment as the CD format and some of their products were affordably priced for a new generation of LP collectors.

A trickle of releases started arriving from small labels. As legal (and other) downloads eroded CD sales, major labels rediscovered LPs as a way of offsetting declining revenues. New and reissued titles now abound. Prices for secondhand LPs have risen, sometimes stratospherically—a mint-condition mono first pressing of *Revolver* or *Kind of Blue* sells for as much as a decent used car. LP manufacturing plants can barely keep up with rising demand; old gear is being pressed back into service.

Today the LP attracts new appreciation from millennials, nostalgia from baby boomers, and ardor from collectors. As this book was going to press, LP sales have continued to rise, along with streaming, while CD and download sales have continued to fall.

Turntable basics

You can play LPs and 45s on a one-box *phonograph*, though such an archaic idea should stay in the past. A phonograph is generally a cheap piece of junk with a lousy tonearm that damages records. The best way to play records while keeping them in good shape is to add a well-maintained turntable to a component system.

Affordable turntables usually come with preinstalled *tonearms* and *phono cartridges* though you may need to shop for those separately when buying a high-end turntable. Unless you have expertise in the art of mating turntables, tonearms, and cartridges, let a good specialty retailer be your guide. If your turntable already comes with tonearm and cartridge, it's probably best to stick with the tonearm, though you may eventually want to upgrade the cartridge.

To interface with your turntable, your system will need either a *phono input* or a *phono preamp*. That's because the phono cartridge puts out a tinier voltage than other audio source components and requires special equalization to preserve the correct balance of frequencies. A suitable phono input may already be built into your receiver, integrated amp, stereo preamp, or (in rare cases) powered speakers. But if your system does not have a phono input, it will need a phono preamp—a separate component that can accommodate a turntable. That may cost a few hundred to a few thousand dollars.

Some inexpensive newbie-friendly turntables have phono preamps built in. You can plug their analog outputs into any of your system's analog line inputs. If the turntable has a USB output as well, you can plug it into your computer, which can be handy when transferring music from vinyl to digital file formats.

Turntable parts

A turntable starts with a *base*, also known as a *plinth* or *chassis*. It also includes a motor, bearing, platter, tonearm, cartridge, and stylus. The heavier the base is, the better it may be at controlling vibration. To guard the platter and tonearm from motor vibration, step-up models may have a *floating top plate* or a *subchassis* suspended from springs.

The motor may be connected to the platter via either belt drive or direct drive. *Belt drive* uses a big rubber band that wraps around both the platter and a pulley connected to the motor. This is the predominant audiophile choice; it is believed to be the best way of further guarding the platter from motor vibration.

The alternative is *direct drive*, which connects the platter directly to the motor. Direct drive is easier to maintain because there is no belt to wear out and replace. Direct drive is also more rugged, and is therefore the preference of hiphop DJs. It is not impossible for an audiophile-quality turntable to use direct drive. However, it must be carefully implemented to prevent the motor from smearing the sound. Some direct-drive turntables are also prone to *hunting* for the correct pitch, audibly wobbling back and forth.

Turntables can be manual, semi-automatic, or fully automatic. This refers to the handling of the tonearm. Most turntables nowadays are *manual*. They require you to set the tonearm onto the record and lift it at the end of the side (or whenever you're done). A *semi-automatic* turntable requires manual operation to start but has a mechanism that automatically picks up the tonearm at the end of the side. A *fully automatic* turntable has a mechanism that starts and finishes play. Automatic turntables have more parts and more ways to go wrong. Audiophiles mostly avoid them because the added mechanical complexity doesn't do anything to improve sound.

If you plan to play both LPs and singles, you will need to adjust *rotational speed*. Some turntables make this simple with an easily accessible 33/45 rpm *speed control*. Others require you to adjust speed by moving the belt from one pulley to another, which in turn may require you to lift the platter from the spindle.

In addition to the nominal rotational speed, you may also need to dial in the actual rotational speed—in other words, to make sure your LPs are spinning at exactly 33-1/3 rpm (revolutions per minute), as opposed to something marginally faster or slower. A different kind of speed control enables this fine-tuning and you might prefer it to be easily accessible. A *strobe* may accompany the speed control, enabling you to find the correct speed at a glance. If your turntable does not come with a strobe, you can assess speed by dropping a downloadable paper strobe disc on top of the platter.

The *platter*, of course, holds the spinning record. It may be made of metal, glass, acrylic, ceramic, or other materials. The manufacturer should have something to say about why its choice of material is best and how it fits in with the rest of the design.

The platter usually comes with a *mat*, which you can replace if you think you can improve on it. Mats may be rubber, cork, or some kind of fiber, each of which has its own sonic properties. Fiber mats may be dust magnets, which may be bad for your records. My beloved Micro Seiki turntables have rubber mats, which render their aluminum platters acoustically inert. Otherwise the metal platters would ring like a bell.

A turntable base also includes *feet*, a seemingly small but acoustically meaningful detail. Better turntable designers choose feet for their vibration-controlling properties.

Most mass-market turntables also include *dustcovers*. Some audiophiles omit them, claiming that the rotation of the platter draws dust onto the record. My experience suggests that a spinning record attracts less dust when covered and a covered turntable is less likely to get filthy.

Tonearms

You probably know that a turntable needs a tonearm to hold the phono cartridge, and a phono cartridge to extract information from the record grooves. What you may not know is that your choices of tonearm and cartridge are vital to the turntable's performance. They contribute as much to the sound as the turntable itself.

In an affordably priced product, the choices of tonearm and cartridge are made for you. But if you are buying a high-end turntable, you may have to pay more attention to these details. It is difficult, though not impossible, to upgrade a tonearm—if you're not a DIY type, a good dealer can do it for you. The tonearm needs to be a certain size for the turntable and for the *armboard* into which it is mounted (unless you're replacing that too). Swapping out cartridges is a bit easier as long as you have a gentle hand and a talent for dealing with tiny screws and wire clips. Otherwise, get it done for you.

A tonearm may be *straight* or *curved*. These shapes offset tracking error in different ways. Most tonearms are metal though some, like the legendary Infinity Black Widow, are resonance-resistant carbon fiber. Some newer carbon-fiber tonearms are 3D-printed.

The tonearm has a *phono cartridge* mounted at the front end and a *counterweight* mounted at the back end. The counter-weight is movable to adjust *stylus pressure* (or *tracking force*). The tonearm will also include an *anti-skating control* to offset forces that pull the stylus toward the center of the disc.

Even the simplest turntable needs to be set up with the correct stylus pressure and anti-skating. They should be set to the same numerical value, in grams. Most phono cartridges work best between one and two grams (look up the manufacturer's recommendation in the manual).

There are other adjustments that, properly done, would improve the performance of your turntable and cartridge. However, you may need expert help to get them done right.

One such adjustment is *VTA* or *vertical tracking angle*, which governs the relationship between stylus and groove. That means raising or lowering the back of the tonearm to get the bottom of the cartridge body parallel to the record.

Overhang adjustment correctly positions the cartridge as it slides up or down the *headshell*. The cartridge manufacturer may provide a paper *protractor* to get it right. That one is fairly easy.

Finally, *azimuth adjustment* is intended to keep the *cantilever*—the part of the cartridge that holds the stylus—perpendicular to the record surface. This may require shims to be installed between cartridge and headshell, almost certainly a job for your dealer's technician.

The tonearm will have a *cueing lever* to enable you to set down the tonearm on the record. Unless you have an exceptionally steady hand, you should use the cueing mechanism when playing records to avoid damaging them. A good cueing mechanism is *viscous*, which means it sets down the tonearm slowly. If the action is too fast, set down the tonearm more gently.

Phono cartridges

Like the tonearm and the turntable itself, the phono cartridge significantly affects the performance of your analog rig. If you know what you're doing, you can improve the sound by changing the cartridge.

The cartridge fits into a *headshell* at the front end of the tonearm. The headshell may be built into the tonearm and inseparable from it, or it may be detachable, which makes mounting a cartridge easier. A *P-mount* cartridge fits directly into a P-mount tonearm, which makes it easy to replace, though this limits your choice of cartridges to P-mount models.

There are two basic types of phono cartridge: moving-magnet and moving-coil. Both produce a signal using a combination of magnets and coils. The main difference is which moves and which is fixed.

In *moving-magnet* cartridges, the magnets move and the coils are fixed. MM cartridges cost less, produce a more robust output that needs less amplification from the phono input or preamp, sometimes come with easily replaceable plug-in styli, and are usually the only type supported by the phono inputs on less expensive receivers and integrated amps.

In *moving-coil* cartridges, the coils move while the magnets are fixed. This produces lower output which needs more amplification, but audiophiles say it also affords finer-grained high-frequency response. MC cartridges usually cost more and require a compatible phono input or phono preamp. Some phono inputs and preamps, especially on more expensive gear, have separate settings for MM and MC cartridges.

The most important part of the phono cartridge is the *stylus*, which is generally diamond, though lower-quality materials may be used in the cheapest gear. The styli in more affordable MM cartridges are diamond *bonded* to steel; better ones are *nude* pieces of diamond. Styli can take different shapes. In ascending order of tracking ability, which may translate to better sound quality, stylus tips may be *conical, elliptical,* or more exotic shapes such as *Shibata* or *micro-ridge.*

Styli wear out, generally after 500 to 1000 hours of use. If you play multiple records every day, you should replace the stylus at least once a year to avoid damage to your vinyl. If your cartridge has a *plug-in stylus,* you can easily replace it yourself. Otherwise you would have to either replace the cartridge altogether or pay to have it *retipped,* which might be worth it if you invested in a pricey cartridge. Eventually the other parts wear out too; no cartridge lasts forever. But a good cartridge should last several years and survive at least a few stylus replacements or retips.

Interpreting turntable specs

A few specs may guide a knowledgable turntable purchase.

The key spec is *wow and flutter,* both of which refer to small variations in speed caused by various mechanical and electronic factors. Slower variations, below 10 per second, cause *wow,* which sounds like what the word suggests. Faster variations pro-

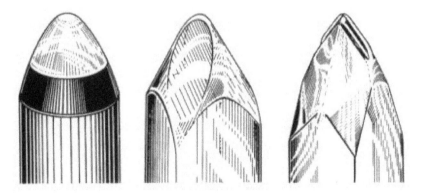

Stylus shapes, from left to right: conical, elliptical, micro-ridge. The more exotic shapes may track better.

duce *flutter*, more of a tremble. Wow can be caused by an LP with an off-center spindle hole but usually it's the fault of the turntable. Both wow and flutter are most easily heard in long steady tones such as a piano note. People differ in their sensitivities to wow and flutter, but once you've learned to recognize them, they can become quite annoying. Wow and flutter are specified as a fraction of a percent. The lower, the better. A wow-and-flutter spec of 0.05 percent is excellent though good turntables, especially affordable ones, may go higher.

Inexpensive turntables may have *platter wobble*. This is especially true of cheap 'tables with thin pressed-metal platters, which are often warped. If you watch the bottom edge of the platter, you may see it moving up and down. This would make it harder for the tonearm and cartridge to track the record accurately. If the wobble is severe, it may add to wow, and a warped record would compound the effect. A well-made turntable should not have a wobbly platter.

Speed accuracy is desirable. You want your LPs spinning at precisely 33-1/3 rpm and your singles precisely at 45 rpm. Many turntables have less than perfect rotational speed out of the box but most can be adjusted. The more often a turntable's rotational speed drifts, the more desirable an *easily accessible* speed control is. If you have to lift the platter to get at it, you'll be annoyed. Most listeners don't notice minor deviations from correct speed but that depends on your sense of pitch.

Interpreting phono cartridge specs

Your choice of phono cartridge will affect the sound of your analog rig as much as your choice of turntable. With phono cartridges, as with speakers, *frequency response* is an important spec, revealing the cartridge's distribution of highs and lows. Different cartridge makes and models can have wide tonal variations, so pay attention to this. Bass may be solid or lightweight, fat or lean, clear or congested. Midrange variations will affect the timbre of voices. The top end may be extended or muffled, comfortable or fatiguing. If you can't demo a cartridge before buying, read as many reviews and user reviews as you can.

Other cartridge specs: *Channel balance* is the variation between left and right channel levels, specified in decibels (dB). A well-made cartridge should not noticeably favor one or the other, so the lower, the better. A channel balance spec of less than 2dB is about average. *Channel separation* refers to leakage between channels, or crosstalk. It is also specified in decibels, but at far higher numbers with 25dB being about average. (Digital audio has far more consistent channel separation.)

Recommended tracking force is the stylus pressure recommended by the cartridge manufacturer. Between one and two grams is optimum. Lower tracking force does not necessarily mean less record wear—a stylus that does not ride securely in the groove may damage the record. If you are having tracking problems, even though tracking force and anti-skating are correctly set, you may want to try higher settings, up to the top end of the manufacturer recommendation.

Output voltage refers to the electrical signal the cartridge produces from its magnets and coils. If you compare one cartridge to another, the one with the higher output voltage will sound louder. Don't worry too much about this. You can generally assume that a moving-magnet phono input (or phono pre-amp) will accommodate the range of voltages produced by most moving-magnet cartridges. Ditto with moving-coil equipment, though there is more variation. Some phono preamps are adjustable to produce better results from a wider range of cartridges.

Finally—and this might seem fiendishly complicated—keep in mind that a good mating of tonearm and cartridge depends on the tonearm's *effective mass* (the mass of its moving parts) and the cartridge's *compliance* (the stiffness of the cantilever which holds the stylus). An effective mating of tonearm and cartridge is one that limits sound-polluting resonance. Generally a low-mass arm requires a high-compliance cartridge. If you'd like to research mass and compliance specs and figure out an appropriate match for yourself, see Galen Carol's helpful tutorial at gcaudio.com/tips-tricks/tonearm-cartridge-compatability (note misspelling of compatibility). Otherwise, if you'd rather not wrap your mind around these thorny issues, an experienced dealer should know what mates well with what.

Cables

Cables are necessary but controversial. Some argue that premium cables are a crucial system component and can make your system sound better; others say expensive cable is a cynical attempt to part you from your money. Early in my career as an audio critic, I experimented with different kinds of cable. I found that speaker cables are the kind most likely to have audible differences, followed by analog interconnect cables, followed (distantly) by digital cables such as coaxial, optical, and HDMI.

So I suggest choosing at least your speaker and analog interconnect cables with care. I'm not telling you to spend more on your cables than you do on your components. But even with a starter system, the best cable for you may not necessarily be the very cheapest. It may be only slightly more expensive, though. And step-up systems deserve step-up cable, up to a certain point.

Speaker cables

Raging controversies notwithstanding, nearly everyone agrees that the thickness of speaker cable can cause an audible difference and that thicker cables are better at minimizing signal loss over long runs. Speaker-cable thickness is specified in *gauge* or *AWG* (*American wire gauge*). The lower the number, the thicker the cable. Default options are 16 AWG for short runs and 12 AWG for longer runs. Avoid cables thinner than 16 AWG unless tiny speaker terminals leave no other option.

Speaker cables use different kinds of *termination*, which seals the copper wire and protects it from oxidizing and losing conductivity. Flexing *banana plugs* are the most convenient choice—as a reviewer, I use them whenever possible. Some prefer U-shaped *spade lugs* because they may provide a larger area of surface contact, though they can slip off unless you use a wrench and extreme care. Some speaker terminals (like those on receivers) won't accept them. *Pin plugs* can fit wire-clip speaker terminals, as an alternative to bare wire, but provide the smallest area of surface contact. If you're stuck with wire clips, you might be better off using bare wire and restripping it once in a while, when the copper darkens. An alternative would be soldered tips.

The fine points of cable construction are beyond the scope of this book. However, here are just a few items: *Copper conductors* are specified in terms of their purity; *oxygen-free copper* (*OFC*) and *silver* strands or coatings are higher-end alternatives. A premium cable may have better *insulation*, to provide protection from abrasions; and *shielding*, to protect against *EMI*, or *electromagnetic interference*, and *RFI*, or *radio frequency interference*. A dealer who lets you borrow cables and listen to them at home may be a valuable ally.

One cheap way to step up to slightly better speaker cable is to choose one with an exterior jacket for better insulation. Such cables can cost barely more than the very cheapest.

Interconnect cables

The same design and construction details that cause audible differences between speaker cables can do the same for *analog interconnect cables*. They can affect the performance of your turntable and any other source component using an analog connection.

There are two kinds of *digital interconnect cables* in an audio system. The *coaxial* type uses the same RCA-type plug as an analog interconnect. The *optical* (or *Toslink*) type usually uses a plastic fiber-optic filament. High-end digital gear may use an optical cable with a glass filament. The back panel of your receiver may label these digital connections as *S/PDIF*, probably in a cynical attempt to mess with your head. It stands for *Sony/Philips Digital Interface*, so let's blame them for it.

Some contend that digital coaxial cables are audibly superior to digital optical cables. If there is a difference, I have yet to hear it, and I've listened for it through some fairly good gear. However, it is undeniable that coaxial cables are sturdier and have a higher bandwidth. Optical cables are more fragile—if you bend them too sharply, they stop working—but they are handily immune from EMI and ground-loop hum.

A good *USB cable* is crucial for the link from computer to USB DAC. A bad one can cause audible glitches. I once reviewed a USB DAC that was plagued with dropouts until I changed the cheap generic cable to a slightly better generic cable with gold tips and a ferrite core filter.

HDMI cables carry both video and surround sound in a home theater system. Better ones carry 18 gigabits per second to support Ultra HD (or 4K) TV. HDMI 2.1, the latest version of the interface, requires a cable carrying 48 Gbps, though you won't need it unless you have an 8K video display (few people do).

Accessories

Do you really need to budget for accessories? You may think of them as things to buy later if you buy them at all. But without certain accessories, your system won't work as well as you'd hoped and you'd be throwing away some of the performance for which you paid when you bought the main components. So plan to spend a few more bucks to get your system to the finish line.

Speaker stands

Unlike floorstanding speakers, which ought to be designed to fire at the correct height, smaller speakers—meaning monitors or satellites—require stands to support them at the correct height. That enables the tweeters to fire at ear level, providing on-axis response, which maximizes performance. It may seem expedient to put speakers on a shelf—aren't some of them called "bookshelf speakers," after all? But a shelf is not the best place for them because it allows undesirable acoustic interaction with the wall, which can muddy the sound. They will sound better a foot or two from the wall.

Speaker stands may have a *resonant frequency* that kicks in when they are vibrated by speakers. In other words, they have a note of their own, and that note is audible when you rap on them with your knuckles. You can definitely hear it with hollow metal stands. To prevent that note from affecting the musically significant midrange, muffle it by filling the columns of the hollow stand with sand or cat litter.

How the bottom of the stand meets the floor also affects sound quality. You have two choices, spikes or feet. Metal *spikes* couple the stand to the floor, drawing resonance from the speaker cabinet, which can produce cleaner sound. And they conduct bass through the floor, which can add a little excitement. However, they can damage wooden floors or disturb your downstairs neighbors. If you're concerned about any of that, use rubber *feet* instead.

Speaker mounts

While most speakers sound better away from the wall, some—including satellites and soundbars—are designed for wall mounting. There are two kinds of speaker mounts: keyhole and threaded insert.

The *keyhole mount* is simplest. Just sink a long nail into the wall and hang the speaker. If the speaker has a *threaded insert*, use a mount that screws into it. You are not necessarily limited to the mount sold by the speaker manufacturer. However, if you buy a third-party mount, be sure the thread sizes match.

Make sure the mount's *rated weight capacity* will support the weight of the speaker. For safety and security, be certain to sink the nail into a wooden stud behind the drywall or plaster. You can find the stud with a stud finder from the hardware store. If a stud is not in the required location, anchor the speaker with a butterfly bolt, which would be less likely to slip out of the wall.

If you rent your home, stands would be a less invasive (and lease-breaking) option than mounts.

Equipment racks

Audio components, especially amplifiers, generate heat. Stacking them and blocking their ventilation holes is always a bad idea. Let them live on a rack so they can breathe. This will prevent them from overheating and shutting down, reduce fire hazards, and add to the longevity of the components.

Shelves may be fiberboard, metal, or glass. If you don't want to find your components on the floor covered in broken glass, avoid glass racks. They have been known to spontaneously shatter due to temperature changes, even when holding less than their specified weights.

Cable management and built-in lighting are among the features you might shop for in a rack. However, the most important things to look for are physical stability and enough shelves to accommodate all your components with little or no stacking. Some racks have modular designs that enable you to add as many shelves as you need. I use, and happily recommend, the Sanus AFA because it is rock-solid, modular, and reasonably priced.

Surge suppressors and power-line conditioners

Don't use a cheap power strip for your system. It will do nothing to protect your gear and may be so flimsily constructed as to be a fire hazard. Instead, get at least a decent surge suppressor, and if possible, step up to a good power-line conditioner.

A *surge suppressor* protects your equipment against surges, spikes, or transients in the power supply. Look for *UL 1449* certification, which indicates that the product has been tested by Underwriters Laboratories with 6000 volts of electricity and will not burst into flames if overheated.

Better products also have *UL Standard 6500* certification which ensures compliance with the safety requirements of the National Electrical Code. You can never be too safe.

The *joule rating* shows how much electricity the surge suppressor can absorb before it fails. Look for four-figure numbers, but take them with a grain of salt because manufacturers make exaggerated claims.

So look for an alternative rating called *voltage let-through*, a measure of how much voltage the surge suppressor will conduct before it clamps down. The less, the better.

Also look for *response time,* a measure of how quickly the product can react to protect your gear. A response time of one nanosecond or less is good. A nanosecond is one-billionth of a second.

Surge suppressors can protect your gear but don't make it sound better. For that, step up to a *power-line conditioner.* Good ones filter out radio-frequency and electromagnetic interference, isolate components from one another to prevent ground-loop hum, and have status indicators for polarity and ground.

Remote controls

If your system has a lot of remote controls and you're tired of switching among them, you may be able to consolidate all their functions into a universal remote control.

Universal remotes come in two types: *preprogrammed* or *learning.* A preprogrammed remote has a built-in library of codes. A learning remote can acquire codes from other remotes. Some remotes are a hybrid of the two.

Perhaps the most versatile remotes are the *computer-programmable* ones from Logitech. A web interface lets you program the remote for your own components and the way you use them. Then a USB cable transfers the control codes from computer to remote.

Turntable accessories and apps

The most crucial turntable accessory is a record cleaner. If you're not going to keep your records clean, you might as well stick with digital audio.

A record cleaner does not have to be expensive. My longtime preference is the old D3 version of the Discwasher, a wet-cleaning system whose brush has angled bristles to coax dirt out of grooves. Other brushes, including the newer D4, are not as effective though I still buy the fluid for use with my vintage D3 brushes. I use separate brushes for slightly soiled and extremely soiled records. I wield the brush under a bright light, spinning the disc on an old direct-drive turntable whose heavy platter moves freely when powered down. To keep the brush clean, I wipe it on a piece of corduroy fabric after each use.

An extremely helpful adjunct to your record-cleaning brush is an *anti-static gun,* such as the one now marketed by Milty. By zapping the static electricity that builds up on records, it loosens dust and makes it easier to remove with a brush.

Record-cleaning machines can salvage extremely dirty LPs. However, if you don't change the fluid and keep the machine itself clean, including the parts that touch the LP, the machine becomes a means of transferring crud from one LP to another. New or well-cared-for used records, cleaned carefully and consistently with a good brush, won't require a machine to rescue them.

In addition to cleaning the record, you should *clean the stylus* after each play. Even one side of an LP can deposit an amazing amount of dirt onto the stylus. Unless you remove it, that dirt gets baked onto the stylus by friction-generated heat and then damages your other records. Use a soft brush, use it gently, and move the brush along the cantilever and stylus toward yourself, never away—you don't want to break off the cantilever. Wipe the brush on something clean after each use. Some wet-system stylus cleaners are available; I haven't tried them. Consistent dry cleaning keeps my styli in good shape. I use an old Watt Dust Bug. You could make something similar by wrapping a piece of velour around a popsicle stick.

Playing a record is a complex series of mechanical events and false resonances can creep in. One way to control them is with a *record clamp,* which fits around the spindle and firmly binds disc to platter. Record clamps can produce audible benefits. However, some heavy clamps may impose more weight on the motor, bearing, belt, or other parts than they were designed to handle. They are not recommended for use with Linn turntables and other turntables with floating sub-chassis because they pull the springs out of balance. To ensure the longevity of your turntable, opt for a lightweight clamp if you use one at all.

A basic kit of *turntable setup tools* should include a small screwdriver to install a cartridge and a pair of needle-nosed pliers to attach tonearm clips to the pins on the back of the cartridge. You'll also need a simple bubble level to ensure that the turntable is level.

A *stylus pressure gauge* can be helpful in setting tracking force though the markings on the counterweight should be accurate enough for that purpose. Shure makes an inexpensive mechanical gauge; other manufacturers offer digital versions.

An anti-static gun like the Milty ZeroStat 3 ($99) greatly improves your record-cleaning routine by loosening dirt before you brush it off. It is well worth the price because extends the longevity and listenability of your records.

To align a cartridge, use the *paper protractor* provided by the cartridge manufacturer. If one is not provided, you can download protractors from vinylengine.com/cartridge-alignment-protractors.shtml. The Mobile Fidelity Geo-Disc is another helpful alignment tool.

If your turntable does not come with a workable strobe, you can measure rotational speed with *phone apps* such as Platter-Speed, RPM, or Turntabulator. Some people claim they are not as accurate as a spinning mechanical strobe. The weight of the phone on the platter might cause a slight speed variation. Paper strobes are available at vinylengine.com/strobe-discs.shtml.

An excellent turntable setup tutorial is *21st Century Vinyl: Michael Fremer's Practical Guide to Turntable Set-Up*. Many other tutorials are available on YouTube. But Fremer's DVD is more authoritative and puts everything you need together in one place.

Recommended Systems

The systems below cover compact stereo systems, component stereo systems, and component surround systems in ascending order of price. Most readers of this beginner-friendly book will find what they want among the first few picks in each category; the pricier stuff is for scale, daydreaming, or when you hit the lottery.

Whenever possible I list equipment about which I have written hands-on reviews. In other instances I list products similar to what I've reviewed, products reviewed by colleagues I trust, or products I've heard demoed at trade shows over the years. In all cases the amplifiers are capable, and often more than capable, of driving the speakers—that is job one when mating audio components. Most of the systems do not include source components but a brief list of those follows. I threw in a few cables as well.

Note that these capsule reviews are written in a few broad strokes; they exclude a lot of relevant information. Visit the manufacturer websites for detailed specifications.

Compact stereo systems

Powered computer or TV speakers: *Audioengine A2+* ($249/pair). The tweeter's top-end rolloff makes harsh computer audio more palatable. It does the same for TV, making these speakers a viable alternative to a soundbar.

The 2.75-inch woofer has a surprising kick for its size—it will sound less boomy away from the wall. Also includes PC-friendly USB and analog inputs (Bluetooth is $189 accessory). Available in black, white, or fire-engine red. See audioengineusa.com.

Stereo system with powered speakers: *Kanto YU6* ($480/pair). If you want substantial sound without a bulky amplifier, the 5.25-inch woofers of these self-powered Canadian-designed loudspeakers play surprisingly deep and loud. An added plus: You can plug your turntable directly into the phono input. The phono input can be switched to a line input (for non-phono sources). Also includes Bluetooth, digital optical, and analog mini-jack inputs. Sold in numerous colors and finishes. See kantoaudio.com.

Stereo system with satellite/subwoofer set and receiver: *Cambridge Audio Minx* ($750), *Yamaha R-N303* ($250). The Cambridge Minx satellite uses a flat membrane to disperse sound evenly around the room, avoiding the beamy quality of conventional cone and dome speakers. Available in single-cube (*Min 12*, $150/pair) or dual-cube (*Min 22*, $300/pair) versions,

the latter with an extra low-frequency driver—but the sats are small, so the complementary sub is a must (*Minx X201*, 6.5-inch, $350; or *Minx X301*, 8-inch, $450). Mate with Yamaha's step-up stereo receiver, loaded with wireless and other features such as Bluetooth, AirPlay, support for numerous music streaming services, phono input, multi-room distribution featuring Yamaha's proprietary MusicCast technology, and a choice of Alexa, app, or old-school remote control. See cambridgeaudio.com, usa.yamaha.com.

Stereo system with mini-monitors and network receiver: *KEF LS50* ($1300), *Marantz M-CR611* ($700). This widely acclaimed little speaker has what KEF calls a Uni-Q array, with the tweeter mounted in the center of the woofer, for pinpoint-clear imaging. Mate with the Marantz network receiver. While this rack-size box may be a stretch for a compact system, its built-in CD drive and triple-threat wireless connectivity (Bluetooth, AirPlay, wi-fi for streaming services) save space by eliminating the need for other source components. For a more stripped-down compact system, a self-powered wireless version of the KEF is also available (*LS50W*, $2200). See us.kef.com, us.marantz.com.

Component stereo systems

Stereo system with monitors and receiver: *Atlantic Technology AT-2* ($1800/pair), *Outlaw RR2160* ($799). Atlantic uses a unique and brilliantly engineered series of ducts to produce true extended bass from a 5.25-inch woofer in a chunky stand-mount enclosure. Factor in the cost of stands for best performance. Mate with the Outlaw's beefy and stylish stereo receiver, which includes PC-friendly USB, network audio via ethernet, digital, analog, and phono inputs (Bluetooth is $100 accessory). See atlantictechnology.com, outlawaudio.com.

Stereo system with powered towers and integrated amp: *GoldenEar Technology Triton Three+* ($2500/pair), *Peachtree Decco125 SKY* ($1199). GoldenEar's extensive line of powered

towers includes this updated version of the Triton Three. Its ribbon tweeter disperses the top end widely, covering the room like a master. The speaker also includes a 4.5-inch mid-bass cone, an oval-shaped sub-bass driver, and two side-firing passive radiators to spread the bass around. The sub-bass driver is backed with its own internal 800-watt amp. To provide more than enough power for the other drivers, mate with Peachtree's sleek-looking and energy-efficient 120-watt Class D amp. It can stream wi-fi straight from your phone without going through your router. PC-friendly USB, digital optical, and analog inputs also included. See goldenear.com, peachtreeaudio.com.

Stereo system with towers and integrated amp: *Dynaudio Excite X38* ($3900/pair), *Luxman L-509x* ($9495). Dynaudio of Denmark is famous for the quality of its drivers. It packs a pair

of seven-inch woofers into these 40-inch-tall mini-towers, in addition to a midrange driver and a smooth coated textile tweeter, built into diecast aluminum frames and backed with large magnets. Available in rosewood or walnut veneers or black or white satin finishes. Mate with a Luxman amp that's retro all the way, from its loveable volume meters, to its defeatable bass and treble controls, to its switchable MM/MC phono inputs, to its total lack of digital inputs—so you can plug whatever digital sources you choose into the back panel without polluting the intricate analog amp circuits. It musters 120 watts into 8 Ohms and 220 watts into 4 Ohms (the rated nominal impedance of the Dynaudios). That should power all but the most demanding speakers with the venerable Japanese high-end company's golden sound. See dynaudio.com, luxman.com.

Stereo system with towers and separates: *PSB Imagine T3* ($7498/pair), *McIntosh C2600 preamp* ($7000), *McIntosh MC452 power amp* ($8500). Paul Barton, the distinguished Canadian speaker designer, uses a seven-layer enclosure and two-inch-thick front baffle to tame resonance in a tower that will never leave you starved for bass. Keeping a grip on its trio of seven-inch woofers may require you to keep the speakers away from the wall, use the supplied foam plugs in their ports, and power them with an amp that has some gumption. Enter McIntosh of Binghamton, New York. The overused word iconic truly applies to this brand, whose wiggling-needle meters and blue-and-green cosmetics are beloved emblems of high-end audio. Enjoy the golden tone of tubes without their dynamic limitations by mating McIntosh's tube preamp with its most powerful solid-state amp, rated at 450 watts per channel from 8 Ohms down to 2 Ohms. See psbspeakers.com, mcintoshlabs.com.

Component surround systems

Surround system with satellite/subwoofer set and receiver: *Q Acoustics 3000* ($900), *Sony STR-DN1080* ($600). Britain-based Q avoids a common pitfall of sat/sub sets—a weak sub—with this 5.1-channel system, featuring a dual-driver small-footprint sub and an elegant round-cornered satellite that's voiced to be smooth, sweet, and spacious. Mate this 5.1-channel speaker package with the latest in a distinguished series of affordable Sony receivers with built-in Bluetooth, AirPlay, Chromecast, and streaming services via wi-fi or ethernet. See qacoustics.com, sony.com.

Surround system with monitors and receiver: *ELAC Uni-Fi* Series ($1797), *Denon AVR-X3400H* ($999). Speaker-design genius Andrew Jones packs three drivers into the little *UB5 monitor* ($499/pair), including a tweeter coaxially mounted within the midrange for pinpoint-accurate imaging. Start with two pairs, then add the *UC5 center* ($349) and *SUB3010 subwoofer* ($450), whose app-driven room correction makes it easy to zap your room's bass bloat. Mate with a Denon receiver that supports the company's HEOS multi-room technology. To adapt this 5.1-channel system to 5.1.4 channels for Dolby Atmos height-compatible surround, add four of the *ELAC Debut A4.2* Atmos-enabled speakers ($250/pair) for the height channels, and power them with Denon's nine-channel *AVR-X4400H* ($1599). See elac.com, usa.denon.com.

Surround system with towers, sub-woofer, and receiver: *Polk Signature Series* ($1500), *HSU Research VTF-15H MK2* ($899), *Rotel RAP-1580* ($3800). Polk has been producing well engineered, high value speakers for decades. The Baltimore-based company steps up the styling in this warm-voiced midpriced series. The *S60 tower* ($900/pair) has a trio of 6.5-inch woofers, midrange, and textile tweeter. Add the *S35 center* ($300), with its unorthodox three-inch woofer sextet, and a pair of the *S20 monitors*

($300) as side-surrounds. The Polk towers have excellent bass response, but for extra force, throw in one of HSU's killer 15-inch subs. Mate with Rotel's top-line "surround amplified processor," the closest modern equivalent to the Rotel surround receiver I loved using for eight years, the longest time I've used any receiver. See polkaudio.com, hsuresearch.com, rotel.com.

Surround system with towers and separates: *Revel Concerta2* ($6650), *Integra DRC-R1.1* ($2500), *Parasound A52+* ($2995). Revel's Kevin Voecks is one of the great American speaker designers and the brand's parent company supports him with world-class engineering resources. This 5.1-channel setup combines the Concerta2 *F36 tower* ($2000/pair) for the front left and right

channels with the *C25* ($750) for the center channel, a pair of the *M16 monitors* ($900/pair) for the side-surround channels, and a pair of the *B10 equalized sub* ($1500). Mate with an Integra pre-pro to get the latest features and with a Parasound five-channel amp (from John Curl, another great designer) for deep bass and low noise. For 5.1.4-channel Dolby Atmos height-capable surround, add four *Revel C763L* ceiling-mount speakers ($750/each) and a second Parasound amp. See revelspeakers.com, integrahometheater.com, parasound.com.

Source components

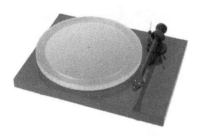

Basic turntable: *Pro-Ject Debut Carbon* ($400): Avoid the very cheapest turntables with their wobbly platters and fluttery sound. This Czech-made belt-drive model features a preinstalled Ortofon 2M Red MM cartridge and a one-piece carbon-fiber tonearm, which is better at defeating false resonance than the aluminum tonearms common in budget turntables. The painted fiberboard base, available in numerous colors, is both a visual and sonic improvement over the piece of plastic junk you almost bought. Pro-Ject 'tables get bigger, better, and more solid as you move up the line. See pro-jectusa.com.

Step-up turntable: *Technics SL-1200GR* ($1699). The return of this classic direct-drive model means that you no longer have to seek out battered vintage versions. The latest versions cost more than the older ones but reviewers say the extra build quality and speed accuracy are worth it. Any version of this beloved turntable, whose enduring place in popular culture was cemented by hiphop DJs of the 1970s, will be rugged and built to last. Cartridge not included. My beloved Shure M97xE is no longer made, so ask your dealer for recommendations. See technics.com.

High-end turntable: *Linn Sondek LP12* ($2630 base price). This longtime audiophile favorite is a closed system, so it's all about the add-ons. For best results you'll

want a Linn tonearm, a Linn cartridge, a Linn phono preamp, and as many of the numerous modifications and upgrades as you can afford. The full kit can raise the price to nearly five figures. Despite the tweak appeal, this Scottish-made classic is soothingly understated in its wood-veneer styling and not nearly as massive as much of its bleeding-edge competition. However, it should not be underestimated. See linn.co.uk.

CD player: *Rotel RCD-1572* ($899). If you're too into CDs to settle for a cheap Blu-ray or DVD player, Rotel offers a CD-only player with one of its sweet-sounding DACs. Extra points for durability—I've got an old Rotel that's been going strong since the late 1980s. See rotel.com.

Media player: *Bluesound Node 2* ($499). Grabs high-resolution audio from your networked computer or mobile device, feeds it to an audio system, and flings it throughout the home using its own wireless protocol. Supports Bluetooth, major streaming services, and integrates

with the major custom-install touchscreen platforms plus the excellent music management software Roon. Step up to the *Powernode 2* ($799) with built-in 60-watt stereo amp—just add speakers—or the *Vault 2* ($1199) with built-in two-terabyte hard drive and CD slot for easy ripping. See bluesound.com.

USB DACs

Note: If you get most of your music from a laptop, a USB DAC (digital-to-analog converter) is not just an accessory—it's an absolute must-have. All of these DACs plug into a PC's USB input, can output to both audio systems and headphones, and are compatible with new and highly advanced MQA encoding.

Basic DAC: *AudioQuest DragonFly Black* ($99). This thumb-drive-shaped device can significantly improve the sound coming from your PC. It can also run off smartphones with an Android OTG adapter or Apple camera adapter, if you can live  with the extra bulk. Like the Drag-onFly Red (see below), the Drag-onFly Black is limited to 24/96 resolution (no 24/192 or DSD). See audioquest.com.

Step-up DAC: The *AudioQuest DragonFly Red* ($199) one-ups the DragonFly Black with a glossy red housing, slightly greater aural refinement, and a smidgen of extra power for slightly more demanding headphones. If the DragonFly Black is like cleaning a smudged window, the DragonFly Red is like opening the window. See audioquest.com.

 High-end DAC: *Meridian Prime* ($999). Designed by MQA co-inventor Bob Stuart and his team, this six-inch-wide box is both USB DAC and analog-input headphone amp. It includes two bypassable listening modes that manipulate the soundstage. To obtain more thorough separation of the analog and digital circuits and therefore purer sound, the

questing audiophile may add the *Prime Power Supply* ($750). Special order through dealer. See meridian-audio.com.

Cables

Basic speaker cable: *Media-bridge* 14AWG ($30/100 feet). Why buy this instead of the cheapest possible zip cord? It has an exterior jacket to protect the leads, properly color-coded red and black leads, and is CL2-certified—meaning fire-resistant and suitable for in-wall use. See amazon.com.

Step-up speaker cable: Canare 4S11 ($59/10-foot pair with banana plugs). Made in Japan and assembled in the U.S., this 14AWG cable uses a quad-conductor design to reduce electromagnetic interference. Unlike a lot of premium cable, it is sold for both pro and consumer use, is reasonably flexible, and is not clad in any kind of wrist-thick snakeskin that would frighten the horses. See chromaleaf.com, canare.com.

Basic analog interconnect: *KabelDirekt* ($9.49/3-foot pair). This German company manufacturing in China makes an affordable but sturdy oxygen-free copper cable with double shielding to beat interference, and metal-housed gold-plated plugs that are secure without being death-grippy. Available in sizes from 3 to 25 feet. See amazon.com.

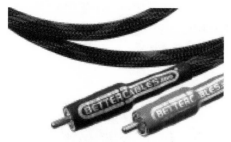

Step-up analog interconnect: *BetterCables Silver Serpent Anniversary Edition* ($68.60/3-foot pair). For more demanding high-end uses, these American-made cables have silver-coated copper conductors, braided shielding, and rugged jackets that can take a beating. I have used them for years in my reference system and they are worth the money. Available in sizes from 1.5 to 49 feet. See bettercables.com.

Product updates on the web

Amazon affiliate and manufacturer links to these products are available on the web: quietriverpress.com/friendly2.html. The book will not receive annual updates but the website will track changing product availability.

Setting Up a System

This brief seven-page guide to setting up a system applies mostly to a relatively simple component stereo system. For a complete guide on the more complex process of setting up a home theater system with a video display and surround receiver, see the 28-page guide in my other book *Practical Home Theater*.

Position components

As you unpack components, store the crates for at least the warranty period of each product. You can save space by flattening the cartons, putting the foam inserts in trash bags, and storing them in a closet. Create a file for instruction manuals and jot the serial number of each component onto the back of each manual. You may need it when calling tech support.

Before setting up your system, give serious thought to how components will be positioned. Don't give in to the temptation to stack them on whatever surface comes to hand. Blocking their ventilation holes would be a potentially fatal mistake, especially with a receiver or amplifier, which generates more heat than most other components. Use a rack to ensure proper ventilation, to avoid a fire hazard, to keep front-panel controls accessible, and to minimize back-panel cable distances.

The most common place for the rack is between the speakers. This is the best way to minimize speaker-cable distances, thus preventing signal loss, which would weaken sound. An alternative place for the rack would be on a side wall, which would require longer speaker cables of equal length. This would cause a minor amount of signal loss, but as long as your speaker cables are not several dozen feet long, side-wall placement is not a bad option. Use thicker 12 AWG cable for longer speaker-cable runs.

Position speakers

Where you place your speakers will have a profound impact on how they sound. The most convenient place—for most people, that means up against the wall—is rarely the best. First consult the speaker manual to see what the manufacturer recommends. Start with that. Then do some fine-tuning.

Most speakers sound their best anywhere from one to several feet from the wall. By minimizing undesirable acoustic interaction with the wall, this results in a clearer midrange and less congested bass. If you want more bass, move the speakers closer to the wall and/or corners. However, more bass is not necessarily better bass. A balanced sound with clean, even bass response is the most listenable in the long run.

Finding the best speaker placement involves moving not just the speakers, but the listener. Sitting too close to the back wall can also muddy bass response. Move the seating away from the back wall, closer to the center of the room.

The distance between a pair of stereo speakers, or the front left and right speakers in a surround system, will affect the *stereo soundstage* and the *imaging* (reproduction and placement) of objects between the speakers. If the speakers are too close together, the soundstage may be too narrow. If they are too far apart, the soundstage may collapse. Start with the two speakers and the listening position in an equilateral triangle. Then try moving the speakers a little closer together. Seek the best compromise between soundstage width and image density.

Another way to improve imaging is to *toe in* the speakers, angling them toward the listening position. This maximizes the proportion of sound heard on-axis. If the highs are fatiguing, or if you would prefer a larger sweet spot, *toe out* the speakers a little, though they still don't have to be parallel with the walls. A good toe-out position lets you see a just a little of the inside cabinet walls (right side of the left speaker and left side of the right speaker).

By coupling speakers to the floor, *speaker spikes* can improve bass extension and control. However, they can make a mess of your floors when you're experimenting with speaker placement. Install them after you are confident that the speakers are where you want them to be.

In a surround system, try to keep the center speaker at the same level as the front left and right so that lateral pans won't waver up and down. Angle it up if necessary. Side-surrounds go on the side walls, firing at the seating position. Back-surrounds go on the back wall, behind the seats. The subwoofer should not necessarily go in the corner—as with speakers in general, more bass is not necessarily better bass. Try various positions between the front speakers or between the front and rear speakers.

Connect speakers

Follow the standard color coding on speaker terminals at both ends, speaker and amp. Red always goes to red and black always goes to black. The red terminals are the positive (+) or hot connection, and the black terminals are the negative (-) or ground connection. If your speaker-cable plugs or jackets are not color-coded, maintain consistency by looking for a stripe or ridge along one side of the cable.

When a speaker connection is mismatched, the affected speaker will run *out of phase*. In other words, the drivers will move in when they should be moving out, and vice versa. This can result in a bass-cancellation effect audible as hollow, disembodied, unsatisfying bass response. If your first impression of your system is that something is askew, check the speaker connections immediately and correct them if necessary.

As explained in the chapters on speakers and cable, speaker-cable tips may be *terminated* in various kinds of hardware to prevent corrosion. Alternatives include banana plugs, spade lugs, pin connectors, soldered tips, and bare tips.

Some speakers come with dual sets of binding posts for *bi-amplification* or *biwiring*. Biamping allows the speaker to receive more than one channel of amplification. That can improve dynamics, making your system play louder and clearer. Biamping is an option with some seven-channel (or more) surround receivers, replacing the back-surround speakers. Biwiring gives each driver a separate path to the same amp; the benefits are more subtle. If you don't want to use these options, just bridge the two sets of terminals using the hardware provided by the speaker maker. Most biamp/biwire speakers are shipped with the bridges in place. You can connect the speaker cable to either the top or bottom terminals.

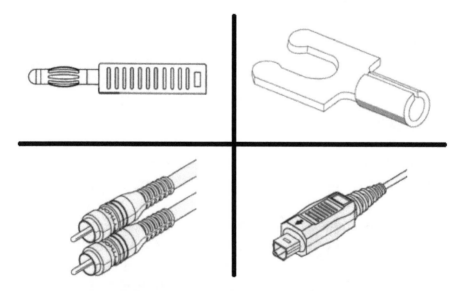

Clockwise from top: Banana-plug speaker-cable termination, spade-lug speaker-cable termination, Toslink (plastic digital optical) interconnect plug, RCA analog interconnect plugs. A single RCA plug would be used for a digital coaxial connection.

Connect source components

When connecting source components—disc players, turntables, etc.—always connect input to output and output to input. Don't connect an input to an input or an output to an output.

Some source components offer more than one connection. The best choice depends on the source component.

Audio-only source components may have digital or analog outputs. If your receiver, integrated amp, or preamp has digital inputs, usually the digital (coaxial or optical) connection is the best choice. However, if the source component features superior digital-to-analog conversion—such as a high-end disc player may have—then the analog connection may be better.

A turntable has special needs. It must be connected to a phono input to ensure that the weak signal produced by the phono cartridge is correctly amplified and equalized for the proper distribution of highs and lows. If your receiver, integrated amp, or preamp does not have a dedicated phono input, you will need to add a separate component called a phono preamp. The phono input or preamp must match the cartridge type, either moving-magnet (MM) or moving-coil (MC).

With audio/video components such as Blu-ray players, cable boxes, DVRs, and streaming boxes, the HDMI connection is best if you are using a surround receiver or preamp-processor. HDMI carries both video and audio. Otherwise you would need to make separate (and possibly lower-quality) video and audio connections.

A component that connects to your home network will have either a wired ethernet or wireless wi-fi connection. Follow the manual's networking instructions.

Bluetooth, on the other hand, is a direct device-to-device connection. Perform the pairing in your mobile device's Bluetooth menu. Some products let you do this with a "bump."

Connect preamp to power amp

If you're using separates—a stereo preamp with a stereo power amp, or a surround pre-pro with a multi-channel power amp—use the three-pin XLR connection if available. It is more resistant to signal loss, especially over long runs, and the third pin can prevent ground hum. Otherwise use the two-pin RCA connection.

Connect power

Only after you've made all other connections should you connect the power cables. To avoid electromagnetic interference, avoid running power cords in parallel with audio cables. Use a good power-line conditioner if possible. Otherwise, at least try to avoid plugging your system into a power line that's shared by an air conditioner. As the AC's compressor switches itself on and off, surges or dips may occur in the power line, jolting your audio components. This would reduce their performance as well as their longevity.

A/V receiver settings

Setting up an a/v receiver or surround receiver is beyond the scope of this book. Be advised that it will require you to handle numerous settings through the onscreen interface. Most receivers have an auto setup and room correction feature. It requires you to connect a setup microphone. Then it spits out test tones to determine things like speaker level, size, and distance, and finally it optimizes sound for the room. For best results, enter your speaker configuration before running the program so that the receiver will be aware of what it's setting up and correcting. And feel free to turn off the room correction if you don't like the effect.

Tweaking

There are many small things you can do to improve system performance. Some are worth considering—but don't let tweaking become an obsession. Tweaks can be acoustic or mechanical.

Acoustic tweaks can make a difference. Your system uses a combination of direct and reflected sound to reproduce music. In a concert hall, most of the sound you hear is reflected. At home, you need to control some of that reflected sound for best results.

Clap your hands and listen to the room. You will hear a *slap echo*. That is a set of reflections off the walls, floor, and ceiling. The louder and more complex it is, the more it is affecting the music. It is not possible, necessary, or even desirable to eliminate the slap echo with draconian measures. But a few basic room treatments can manipulate room reflections to improve sound.

First, floor reflections are never good. Reduce them to the extent possible by getting a rug to cover most or all of the floor. A heavy pad beneath the rug can help.

To find out where sound is reflecting off the side walls, have an assistant move a hand mirror along the walls until you can see the speaker drivers from the listening position. Those side-wall hot spots are good places to put your media-library shelving. Otherwise, absorb or diffuse side-wall reflections with tapestries, wall-hanging rugs, foam rubber, or egg carton.

Back-wall reflections should be treated as well. You can tame them by moving seating away from the back wall. However, more shelving or other absorption/diffusion devices might also help.

One of the most effective ways to tweak your system is to experiment with placement of both speakers and seating. You can never do too much of that.

In addition to acoustic tweaking, you might also fool around with *mechanical tweaking*. Killing mechanical resonance is an audiophile obsession that has spawned a whole genre of accessories. They work in one of two ways.

Energy dampers look like hockey pucks. Place them under components to absorb vibration that might pollute delicate circuits.

Energy couplers are metal cones or spikes that soak up energy on their broad sides and transmit it through their sharp tips. Their primary benefit is to either drain mechanical energy from components, or to prevent floor-borne vibration from affecting components. Speaker spikes are the primary example—you'll like the way they focus bass pitches. But energy couplers may also be placed beneath source components.

Collecting a Music Library

A re hard-copy music libraries obsolete? They take up space and cost money, competing with other household needs and raising the significant other's eyebrow. You could spend a fortune and still barely compete with the vasty deeps of a major streaming service. Streaming is the world's biggest music library.

But just as radio was once the music-discovery medium that drove record sales, streaming now does the same for the contemporary collector. You stream it and then you've got to have it.

Older methods of music discovery still apply: Your friend shares a favorite album and you've got to have it. You hear a song on a movie soundtrack and you've got to have it. Eventually collecting becomes a pastime in itself: you forage in used-LP bins or on Ebay, find something intriguing, and you've got to have it.

Why collect?

Playing hard-copy music formats, especially LPs, is a multimedia experience. That 12-by-12-inch record jacket has been a fertile medium for artists like Roger Dean and design firms like Hipgnosis. The bending of the jacket photo on *Rubber Soul* signified that something trippy was happening with the Beatles. Andy Warhol's zipper jacket for *Sticky Fingers* by the Rolling Stones was pure sex and brilliant marketing. Barry Godber's cover painting for the debut album by King Crimson (inset) was provocative and almost

81

terrifying. Perhaps the most luxuriously tactile gatefold of all time was Gary Burden's leather-textured jacket for Crosby, Stills, Nash & Young's *Déjà Vu*, with gothic lettering stamped in gold, complementing the band portrait in Civil War garb. Reducing this brilliant art to a thumbnail for streaming or downloading robs the listener of visual stimuli that the artist and record label intended to accompany the music.

Streams and downloads lack information routinely provided on disc packaging. Open the gatefold, glance at the back of the jacket, or unfurl the CD sheet or booklet to find out: Who wrote that song? Who's in the band? Who played drums on track three? Who produced the album? Where and when was it recorded? That who-what-where-when provides clues that may be critical to understanding the music (even if the why is something you'll probably have to figure out for yourself).

Collecting LPs

The golden age of LP collecting started with the format's debut, continued with its status as the dominant longform music medium, and ended with a dramatic blowout as people threw out perfectly good records and replaced them with CDs—a glorious opportunity for the budding record collector. Today it is still possible to build a great LP library, but you have to work a little harder and spend a lot more. You are competing with an increasing number of people willing to pay crazy money for vinyl, new or old.

Should you buy new music on LP? Technically, most music is digitally recorded and mixed, so in theory a high-resolution digital medium may be better. If the final master is a 24-bit, 96 kHz file, you might get better sound by simply downloading the file in a 24/96 lossless format. But if the album is not downloadable in a high-res format, the LP release may have higher resolution than the CD release, with the latter automatically downconverted to 16/44.1. A CD is easier to rip for use on your phone, but some labels anticipate that issue by including MP3 download codes with LP releases.

For back catalogue, you have a choice between modern reissues and vintage pressings. The latest remastering may purport to be superior; new mixing technology and/or better ears may produce better results. But reissue producers face hurdles. The master tape may be lost or in bad condition. The reissue may be based on an inferior copy of the master tape. In the worst-case scenario, CD masters—intended for a different format with different technical requirements—are thoughtlessly repurposed for LP releases. Seek out reviews by knowledgable people who are in a position to compare the reissue to the original issue.

The other challenge, even when you're buying either reissues or new releases, is getting a disc with no severe warpage and a clean playing surface with only minimal clicks and pops. In this respect, record collecting today is the same as it ever was. A retailer with a reasonable return policy makes life better. But don't push your luck. Sometimes a whole pressing is lousy.

Vintage pressings can be better. They were made when the master tapes were fresh, frisky, and unharmed by the passage of time. I'd rather have a vintage issue in perfect condition than a modern reissue any day.

Even so, the vintage-vinyl collector faces two challenges. One is finding a copy in collectible condition. In the grading used by LP dealers, anything below mint or near-mint will probably be audibly flawed (note different grades for disc and jacket).

The other challenge is getting as close to the first pressing as possible. This is not a problem for albums that were not pressed in huge numbers, including most classical releases. However, if you're collecting, say, the Beatles, know that you're up against people who know every stamper number. A lot of detailed information is available online but be prepared to descend into the rabbit hole of obsession.

Collecting CDs

The downward arc of the Compact Disc format has made it a present-day collector's bargain. In the mid-1980s a CD sold for about $15. Adjusted for inflation, that is more than 30 twenty-first century bucks. Today many new releases sell for less than $10 and many back-catalogue titles for even less than that. You may find a CD selling for less than the inferior-sounding MP3 alternative. You could buy the CD, rip it to lossless FLAC or ALAC, and end up with better-sounding files.

This is the age of the wallet box. With CD revenues trending downward, record companies are frantically shoveling out their back catalogue in box sets with discs in slim slipcases. Forty discs can fit into a box less than three inches wide. Anyone with shelves full of plastic CD "jewelboxes," the antiquated plastic packaging in which the format made its debut, will appreciate how much shelf space that can save. Wallet boxes are especially common in classical catalogues. With pricing as low as $3 per disc, the wallet box is a boon to collectors.

Beware of CD reissues that substitute CD-R discs for regular stamped discs. While nothing lasts forever, a regular CD (with the digital data stamped onto plastic backed by a metallized reflective layer) will last longer than a recordable CD-R (which uses a dye process). Seek the original stamped CD issue. If you can't get it, downloading is a better alternative as long as it is at least CD-quality (16-bit, 44.1kHz) or higher.

Collecting downloads

While streaming is now outpacing downloading, there are still advantages to owning (as opposed to renting) digital audio. Your library doesn't vanish when you stop paying the streaming bill. You can bring music into places where your ISP can't follow.

My advice on downloading is minimal: Avoid lossy formats such as MP3 or AAC because they may not sound good. Avoid uncompressed formats like WAV and AIFF because they sound no better than lossless formats and are storage hogs. Collect in lossless formats such as FLAC or ALAC because they provide the optimum combination of sound quality and storage efficiency.

For the highest resolution, when you have options, choose 24-bit over 16-bit, with the highest possible sampling rate.

Collecting mania

Collecting has its dark side. Some people collect with little or no intention of listening to what they collect. Unlike streaming and downloading, the population of LPs and CDs in the wild is a zero sum game, at least where out-of-print items are concerned. If you have an item, someone else can't have it—someone who might actually get some musical pleasure out of it.

Of course, if you have a large library, you will rarely if ever traverse the whole thing. I have a library of about 5500 albums in various formats. If I listened to one a day, 365 days a year, it would take me 15 years to get through the whole thing. One of the joys of owning a large library is the freedom to dive into it serendipitously. My criterion for keeping an LP or CD on the shelf is that I might conceivably want to listen to it again some-day. If not, I sell it or (more often) give it away.

But unless you're a record dealer, hoarding discs just for the sake of it can be a sickness. There is not necessarily more music in your life—just more *stuff*. The central question of record col-lecting, as with audio in general, is this: Does it bring you closer to music?

Listening

Would it be unfriendly to conclude *The Friendly Audio Guide* with a chapter on listening? If you've gone far enough to buy a book, you probably know how you like to listen and don't want to be lectured about it. People listen to music in different and often highly personal ways. Mine is not necessarily better than yours; I don't mean to condescend. But I would suggest that the kind of listening that gets you closer to music has become something of a lost art.

There is background listening and foreground listening. I'm not suggesting that background listening is automatically debased. I do a fair amount of it myself. In a busy life, background listening may become the difference between having music or not having it. Audio gear in general has evolved so that you can have pretty good sound anywhere in or out of the home. Technology brings music to the places and circumstances where you need it to go.

But if you have any choice in the matter, the best way to get closer to music is to listen actively. Take the things that compete with music and, one by one, set them aside. Let music be the sole focus of your attention. Let it loom larger as you approach. You may find that focused listening makes possible a wider range of emotional and intellectual responses. They are more intense, more complex. Experience the liberation of mono-tasking and let music engulf you.

If you don't have enough time for intense foreground listening, find the time. Start by finding 20 minutes per day, the average length of an LP side. Extend that to a full album as soon as you can. This is your *me time*. Just knowing that you can rely on it for those few minutes a day may reduce the stress in your non-listening life.

But you can have too much of a good thing. The first album is a great idea. The second album may be too. The third album, maybe not so much, especially if that pesky urge to multi-task reasserts itself. Don't forget that the aim is to increase not just the *quantity* of listening, but the *quality* of listening.

Turning an unused bedroom or basement into a listening room can improve your listening life. But if that's not possible, turn whatever space is available into a listening room. With good speaker placement, a good chair, and maybe a little acoustic treatment, any room can become a listening room.

Make yourself comfortable. Set up a really cushy armchair in the sweet spot—consider it another component of the system. If the seating isn't comfortable enough when you're doing nothing, then it won't be comfortable enough when you're listening.

It is not impossible for a listening room to accommodate other members of the household. It can include a large video display and a surround sound system for family movie nights. But for solo listening, it should also include at least two great speakers, a solid amp, and a sweet spot with the kind of seating that enables you to relax.

This 1980 ad for Maxell audiocassette tape, known as "the blown-away guy," became a symbol of active listening.

To demonstrate the power of focused listening to yourself, try this crafty old audiophile trick. Pick a lazy evening when you have time to yourself. Definitely make it after sundown. Start playing a long piece of music with normal room lighting—let's say, bright enough to read by. Halfway through, dim the lights to almost nothing. Listen in near-darkness. Has anything changed?

Most people find that the music has become more intense. But it is not the music that has changed—it's the quality of listening. When your brain devotes less processing power to what you see, it can devote more to what you hear, and the difference is noticeable. When you listen, you are not using only your ears. You are using your mind as well.

By the way, this mind tweak also works amazingly well at concerts, especially with unamplified music. Just close your eyes, let your brain redistribute your attention, and you'll notice a difference in the quality of listening.

The best way to get closer to music is to do nothing, absolutely nothing, except listen to it. That's the last piece of advice in this *Friendly Audio Guide*.

Acknowledgments

First and foremost, thanks to Henry Finke—who has worked for JBL, KEF, Genelec, and other distinguished audio brands—for being the main expert reader for this book as well as a loyal and tolerant friend for many decades. Thanks also to Brent Butterworth, who helped with my first book, thus indirectly enriching this one. Any remaining mistakes are my own.

Thanks to my editors-in-chief at *Home Theater* and *Sound & Vision*, who supervised my work as an audio equipment reviewer—starting with Maureen Jenson, who brought me on board, followed by Shane Buettner and Rob Sabin. Their support made my job possible and their editing make my work better.

My reviews for *HT* and *S&V* were also improved by audio technical editor Mark J. Peterson, who measured the gear I reviewed and vetted my copy for technical accuracy, with logistical help from studio manager John Higgins. Thanks also to other team members: executive editors Claire Lloyd Crowley and Adrienne Maxwell, and art director Heather Dickson. We're all alumni and alumnae now.

Thanks to Kevin Wang of Chromaleaf (distributor of Canare) and Brad Marcus of BetterCables.com for providing samples of speaker and interconnect cables mentioned in the chapter on recommended gear.

Finally, thanks to S, international man of mystery, who observed my book frenzy with his usual compassion and humor.

Mark Fleischmann
New York, June 2018

About the Author

Mark Fleischmann is a New York-based writer specializing in technology and the arts. He was a co-founder of etown.com (1995-2001), in its time the world's most widely read consumer electronics publication, and was its first and longest-serving editor-in-chief. Mark reviewed audio gear for *Sound & Vision* and *Home Theater* magazines for 17 years and was audio editor for 12 of them. He served as audio critic of Rolling Stone for nine years and senior editor of *Video* magazine for five years. He has been a columnist for *Audio Video Interiors, Digital Trends, Premiere,* and *The Village Voice.* He has been both a movie and video critic (for *Entertainment Weekly, Newsday, Video*) and a music critic (*Musician, Spin, Trouser Press*). He also edited the *Trouser Press Collectors' Magazine.* His other books include the annually updated *Practical Home Theater: A Guide to Video and Audio Systems* and *Happy Pig's Hot 100 New York Restaurants.* Find out more about his books at quietriverpress.com and about his career at quietriverpress.com/mfindex.html. Mark still plays his LPs and still feels silly writing about himself in the third person.

Index

CPSIA information can be obtained
at www.ICGtesting.com
Printed in the USA
BVHW03s1645300718
523023BV00003B/247/P